Indigenous Knowledge and Development

Indigenous Knowledge and Development

Livelihoods, Health Experiences, and Medicinal Plant Knowledge in a Mexican Biosphere Reserve

By Elizabeth Anne Olson

LEXINGTON BOOKS
Lanham • Boulder • New York • Toronto • Plymouth, UK

Published by Lexington Books
A wholly owned subsidiary of Rowman & Littlefield
4501 Forbes Boulevard, Suite 200, Lanham, Maryland 20706
www.rowman.com

10 Thornbury Road, Plymouth PL6 7PP, United Kingdom

An earlier version of chapter 4 was published as "Anthropology and Traditional Ecological Knowledge: A Summary of Quantitative Approaches to Traditional Knowledge, Market Participation, and Conservation," in Culture, Agriculture, Food and Environment 35(2): 140-151. Open access journal.

An earlier version of chapter 5 was published as "Community Development and Outsider Intervention in the Context of Protected Areas: Case Studies from the Sierra of Manantlán Biosphere Reserve, Western Mexico," co-authored with Peter R. W. Gerritsen, in Ecological and Environmental Anthropology 6(1): 10-25. Open access journal.

An earlier version of chapter 6 was published as "Notions of Rationality and Value Production in Ecotourism: Case Studies from a Mexican Biosphere Reserve," in Journal of Sustainable Tourism 20(12): 215-233. Reprinted with permission from Taylor and Francis.

British Library Cataloguing in Publication Information Available

Library of Congress Cataloging-in-Publication Data

Olson, Elizabeth, 1979-
Indigenous knowledge and development : livelihoods, health experiences, and medicinal plant knowledge in a Mexican biosphere reserve / Elizabeth Olson.
pages cm
Includes bibliographical references and index.
ISBN 978-0-7391-7663-4 (cloth : alk. paper) -- ISBN 978-0-7391-7664-1 (electronic)
1. Indians of Mexico--Ethnobotany--Mexico--Reserva de la Biosfera Sierra de Manantlán. 2. Indians of Mexico--Medicine--Mexico--Reserva de la Biosfera Sierra de Manantlán. 3. Indigenous peoples--Ecology--Mexico--Reserva de la Biosfera Sierra de Manantlán. 4. Medicinal plants--Mexico--Reserva de la Biosfera Sierra de Manantlán. 5. Traditional ecological knowledge--Mexico--Reserva de la Biosfera Sierra de Manantlán. 6. Reserva de la Biosfera Sierra de Manantlán (Mexico)--Environmental conditions. 7. Reserva de la Biosfera Sierra de Manantlán (Mexico)--Social life and customs. I. Title.
F1219.1.R47O57 2014
972'.62--dc23

2013045902

™ The paper used in this publication meets the minimum requirements of American National Standard for Information Sciences—Permanence of Paper for Printed Library Materials, ANSI/NISO Z39.48-1992.

Contents

List of Figures vii
List of Tables ix
List of Abbreviations xi
Preface xv
Acknowledgments xvii

1 Introduction 1
2 Sierra of Manantlán Biosphere Reserve 21
3 Conservation and Subsistence Strategies 43
4 Market Interaction and Indigenous People 57
5 Governance: Community Decision-Making 71
6 Indigenous People and Development Programs 87
7 Ethnomedical Systems in Mexico 109
8 Medicinal Plant Knowledge in the Reserve 127
9 Final Thoughts 139

Glossary 149
Bibliography 155
Index 169
About the Author 181

Figures

Figure 1.1. Conceptual model of the principle axes. 4

Figure 1.2. Study objectives and main hypothesis. 5

Figure 1.3. Distribution of study sample by age and gender. 11

Figure 2.1. National map of Mexico and the Reserve. 22

Figure 2.2. Regional map of the Reserve and urban centers of influence. 24

Figure 2.3. Social relationships in the Reserve—a global view. 36

Figure 3.1. Map of the main study communities and the designated core zone of the Sierra of Manantlán Biosphere Reserve. 45

Figure 5.1. Dominant relationships between key stakeholders in community-based participatory research projects. 76

Figure 6.1. Typology of tourism. 92

Figure 9.1. Reality of trade-offs in example of La Casa. 146

Tables

Table 2.1. Surface area covered by the principal land and forest classifications in the Reserve in 1993. 23

Table 2.2. Demographic description of sampling frame and national census. 33

Table 2.3. Descriptive data and census data of study sample for all three communities. 34

Table 3.1. Agricultural fields and animal husbandry by study community. 47

Table 5.1. Comparison of households in Zacualpan and Ayotitlán, select characteristics. 74

Table 5.2. Summary comparison of Zacualpan and Ayotitlán attributes. 81

Table 6.1. Summary of characteristics of two ecotourism projects: Color de la Tierra and Cabañas El Terrero. 106

Table 7.1. Childbirths and health care services in the Reserve. 113

Table 7.2. Semi-structured interviewees. 117

Table 8.1. Construction of socioeconomic indices (yes=1, no=0). 131

Table 8.2. Plants listed by community and whole sample. 132

Abbreviations

ANOVA	One-way analysis of variance
CONAFOR	Comisión Nacional Forestal, or National Forestry Commission
CONANP	Comisión Nacional de Áreas Protegidas, or National Commission of Protected Areas
CUCSUR	Universidad de Guadalajara Centro Universitario de la Costa Sur, or South Coast University Campus of the University of Guadalajara
ICDP	Integrated Conservation and Development Project
IMECBIO	Instituto Manantlán de Ecología, Conservación y Biodiversidad, or Manantlán Institute of Ecology and Conservation of Biodiversity
IMSS	Instituto Mexicano del Seguro Social, or Mexican Institute of Social Security

INEGI	Instituto Nacional de Estadística y Geografía, or National Institute of Statistics and Geography
MAB	Man and the Biosphere
MPK	Medicinal Plant Knowledge
PAN	Partido de Acción Nacional, or National Action Party
PCA	Principal Components Analysis
PRI	Partido Revolucionario Institucional, or Institutional Revolutionary Party
PROCODES	Programa de Conservación para el Desarrollo Sostenible, or Conservation Program for Sustainable Development
PRODERS	Programa de Desarrollo Regional Sustentable, or Regional Sustainable Development Program, which is now PROCODES
PROGRESA	Programa de Educación, Salud y Alimentación, or Program for Education, Health and Nutrition
RBSM	Reserva de la Biosfera de Sierra de Manantlán, or Sierra of Manantlán Biosphere Reserve (SMBR)
SEMARNAT	La Secretaría de Medio Ambiente y Recursos Naturales, or Secretary of Environment and Natural Resources

SPSS	A statistical analysis software package, acronym originally referred to Statistical Package for the Social Sciences
TEK	Traditional Ecological Knowledge
UNESCO	United Nations Educational, Scientific and Cultural Organization

Preface

Indigenous people have long struggled to maintain their dignity in a rapidly changing world. For some indigenous peoples, that has meant retaining aspects of their cultural heritage. For others, it has meant creating ideological and physical spaces in modern societies through means of assimilation and differentiation. Often, as in the part of Mexico where I carried out this research, there is a blend of the two: indigenous people seek ways to retain aspects of cultural heritage and pride while demonstrating their competence and contributions in modern society.

Meanwhile, across the globe we are becoming acutely aware of the limited resources and destruction humans have had (and continue to have) on our environments. Protecting environments is not strictly the work of conservation biologists and governments. Pertinent environmental issues range the gamut from littering, over-consumption and waste, use of fossil fuels, deforestation, agricultural practices, and modes of transportation, to name just a few. In each society or community the environmental concerns differ, owing to differences in the local needs, interests, and capacities. For an array of historical and sociopolitical reasons, which I will not unpack here, many of the most precious and biologically diverse environments remaining are places where marginalized indigenous people have, and continue to, live.

We often associate modern technology with the rampant destruction of natural resources and environments—as though traditional forms of exploiting and interacting with environments did not have the magnitude of negative impacts that modern technologies do. In some respects, it is true that traditional modes of exploiting environments did not have nearly as large nor lasting of impacts as our current technologies do. However, because of the delicate (often frail) condition of many biologically diverse habitats, conservation of these areas has become even more complicated and urgent.

There we have it—humans and environments. The implications are vast; our health, economies, and happiness all depend in some way on how we interact with our environments. Because we share this earth and its resources through varying forms of stewardship, *it* does effect all of us. But, coming back to the matter at hand (that is, why study these relationships in a protected area of Mexico), it has become the onus of indigenous peoples to help conserve biodiversity. Amidst their own struggle to succeed in the globalized twenty-first century, these peoples are also implied in efforts to conserve. There are two main reasons we need to keep indigenous peoples at the center of conservation work: (1) we assume that they have specialized knowledge and technology that is adapted to a specific environment and ecosystem; and (2) they are frequently the rightful stewards of places rich in biodiversity.

Sustainable development is a paradigm that was developed in the 1980s to address this situation that many indigenous peoples were facing. The idea is that we can find ways to exploit natural and human resources now that do not compromise the ability of future generations to also benefit from those resources (Anand and Sen 2000). Regardless of my intentions to strictly study the relationships between traditional knowledge of plants and increasing modernity of indigenous people, I could not escape the pervasiveness of sustainable development activities in my fieldsite. I imagine that many of the people I worked with have felt the same way.

To be an object of study is awkward, and so carrying out anthropological research has the added complexity of navigating the inherent discomfort felt by the communities where we work. Ethnographers suggest that our relationships with community members are sincere, meaningful, and lasting. Certainly, I continue to return to my fieldsite and maintain relationships with many of my key informants. Beyond the personal relationships I developed, the relevance of my research projects to local people is a main concern. The research I completed has both academic value and applied value to the people I worked with. My research, presented here, suggests that local people living in a protected area vary greatly in their cultural knowledge from one community to the next. Owing to this variation in cultural knowledge, it behooves those institutions charged with implementing sustainable development programs to give priority to local-level attributes. Armed with scholarly research, these communities have a stronger voice with which to advocate for their own goals.

Acknowledgments

The research presented here is the result of my doctoral dissertation fieldwork, which was supported by the indigenous communities and individuals in the Sierra of Manantlán Biosphere Reserve (Reserve). I am deeply grateful to the Consejo de Mayores in Ayotitlán, along with the communities of Zacualpan, Telcruz, and Cuzalapa. I gratefully acknowledge the efforts of my field assistants in Zacualpan, Telcruz, and Ayotitlán (especially Sergio and Margarita) who made the completion of data collection possible. There were 125 community members who opted in on my seemingly endless packet of questions. *¡Muchas gracias a todos ustedes!* The communities of Ayotitlán, Telcruz, Zacualpan, Cuzalapa, and El Terrero in the Reserve were the setting for my research, and I continue to be a part of these communities as an anthropologist and humanist. I have grown into their communities in different ways, and certainly some were more eager to receive me at first, but all showed my work and me immense respect. I thank these communities for being genuine in their interactions with me.

Many representatives of the Management of the Reserve and the University of Guadalajara, South Coast, facilitated my fieldwork. The representatives and fieldworkers at the Management of the Reserve have helped my research logistically by providing rides and helping communicate with community leaders. I am also indebted to the many amazing researchers at the University of Guadalajara, South Coast, in Autlán, Jalisco. Through the assistance and encouragement of many of those involved with the Institute for Ecology and Biodiversity I was able to formulate my research question, design my study, and carry out the fieldwork. I hope that the presentation of the information here will be in solidarity with the ongoing community-based efforts to enrich the lives of individuals and families in the Reserve.

A variety of academic and professional mentors have encouraged and supported me throughout the process of my research. Dr. Woody Gaines has been a great influence on my development as an anthropologist, along with the entire Anthropology Department at Case Western Reserve University. Additionally,

Dr. Bruce Benz was critical to introducing me to the Sierra of Manantlán Biosphere Reserve and helping me to develop a testable research question that has local significance; he later also helped me to validate the outcomes of my research. Dr. Peter R. W. Gerritsen has been a great teacher, mentor, and colleague. Friends and fellow anthropologists have lent me a hand and been helpful soundboards. Dr. Robbie Davis-Floyd has stood beside me pushing me forward over the years. Dr. Ruth Gomberg-Muñoz played a significant role in getting me to publish this book.

I have had another non-academic audience that is important because they help me reflect on my work critically and force me to make sense: my Olson family. Thank you for always trying to understand my work and never underestimating the significance of my contributions.

I would also like to acknowledge and thank the following journals that have allowed me to reprint some of my prior journal publications in this book: *Ecological and Environmental Anthropology*, *Journal of Sustainable Tourism*, *Conservation and Society*, and *Culture, Agriculture, Food and Environment.* I greatly appreciate the helpful comments and suggestions of anonymous peer reviewers who have contributed to the presentation of my work and assisted me in refining the interpretations of the analysis. Any errors in this book, however, are my own.

Over the years Autlán de la Grana has become my "other home" as I worked in the Manantlán region. Though I have developed many friendships and professional relationships, there is a special group of people who deserve recognition. The Aldacos are more than my friends and neighbors. *No hay palabras para expresar mi gratitud por la familia Aldaco. ¡Los quiero mucho!*

Me siento honrado de ser parte de las vidas de muchas personas en la región de Manantlán quienes tengo en alta estima. Rezo para que mi trabajo es un testigo a su valor y patrimonio.

Chapter 1
Introduction

This book is about traditional knowledge of medicinal plants in marginalized, developing, forested communities in west central Mexico. My appraisal follows the trajectories of three axes: conservation and development, health and livelihood, and medicinal plant knowledge. These axes intersect in more than one manner or aspect of life. Specifically, there are important historical, cultural, political, and economic aspects within and across each of these central themes. The ethnography I construct in this book expands and contracts around the three primary axes as I take into consideration the fullness of each theme and their intersections. My study is an anthropological exploration of the relationships between conservation and sustainable development, socioeconomic profile, health, and medicinal plant knowledge.

The Research Questions

The goal of my study was to understand and describe the local history, political structure, economic and health profiles in some indigenous communities in west central Mexico. This book explains in ethnographic detail the relationships between conservation, health, and livelihood in the Sierra of Manantlán Biosphere Reserve (the Reserve), giving depth to previously unanswered questions. To do this, I narrate the dynamic system of relationships in a protected area where indigenous people live, specifically the cultural juncture where medicinal plant knowledge is expressed. *Protected areas* are locations that receive special protection (normally through laws and policies) because of their unique environmental and/or cultural value. My research posited: to what extent do economic activities and past illness experiences explain the differences in knowledge of medicinal plants held by members of the same cultural group?

Globalization means increasing interconnectedness of indigenous and local people across boundaries—an unfolding process accompanied by transmissions of knowledge and value systems. By focusing on traditional medicinal plant knowledge (a very elaborated cultural sphere), the study I present in the following chapters is situated to help (de)construct and understand the ways that the livelihoods and culture of indigenous people are associated with medicinal plant knowledge.

The research I present here is significant because it shows that, contrary to what might be expected, there is not an obvious, direct, linear relationship between increasing "modernization" or economic development with the loss of traditional knowledge. Economic changes are central components of the ways in which individuals and communities are facing new problems and opportunities in contemporary Mexico. Furthermore, this research helps to address the possible correlations between variation in livelihood and socioeconomic profile, medicinal plant knowledge, and other factors related to sustainable development and management of forest resources.

In this study, I collected data that can be used to compare intracultural variability of traditional knowledge cross-culturally. Methodologically and theoretically, anthropology has lacked a consistent system for collecting data regarding variation of traditional knowledge within cultural groups. I collected data that can be considered in light of previous work in the Reserve, and prior research in other geographical areas that has been collected at the individual, household, and community levels. An important element of rigor in this study research is the value of understanding the heterogeneity of economic activities within and across communities.

Anthropologists and social scientists have some key reasons for our interest in studying cultural knowledge. First, economic activities are a defining characteristic of human experiences, and I delve deeply into the livelihoods of the Reserve in this ethnography. Second, exposure to global worldviews is becoming increasingly significant, and I explore these in depth in three towns in the Reserve. Third, there is a lack of understanding of the individual characteristics contributing to differences in knowledge about forest resources, which my research seeks to clarify. Fourth, health is a central concern in all societies that demands ongoing development and reformation of systems to promote health. Lastly, cultural heritage—the basis of this ethnography—is a valued aspect of human capital.

My research may also have some potential relevance for ongoing economic programs and initiatives with explicit goals of sustainable development and conservation in the Reserve. The sustainable development paradigm promotes use of natural and human resources in ways that do not compromise the ability of future generations to also exploit the resources (Anand and Sen 2000). This paradigm of improving social and economic opportunities is congruent with many conservation goals, and thus sustainable development has become a guiding theme in the management of protected areas, especially in Mexico (Halffter 1995).

Economic and social "development" may lead to unknown or unpredictable changes in cultural knowledge systems. The research findings I lay out in the coming chapters will help advance knowledge both in and of the communities of the Reserve. If we know what types of economic "development" do not lead to loss of traditional knowledge, than we can advocate for policies and practices that promote not only sustainable economic development, but retain cultural heritage as well.

Study Variables

Ethnomedical knowledge includes allopathic and traditional systems, namely all the methods used to maintain and restore health. I restricted my research to "medicinal plant knowledge" (MPK) referring to the local botany and botanical products that are used in the traditional treatment of illness and promotion of health. The term *medicinal plant knowledge* is unique to my book; in related studies it is common to refer to "traditional ecological knowledge" as TEK, from which I am distinguishing the type of knowledge I studied. Related terms such as *traditional knowledge* and *cultural knowledge* are often used interchangeably to refer to the broad rubric of knowledge and ways of knowing held by cultural groups and transmitted through tradition. I refer simply to MPK, which is synonymous to ethnomedical botanical knowledge, local medical plant knowledge, or indigenous botanical medical knowledge.

I limited my study of medicinal plant knowledge to botanical products and did not seek to identify what people do or report the capacity, which would include practical and skills dimensions (Reyes-Garcia et al. 2006). The research I present herein does not address traditional medicinal knowledge related beyond the knowledge of plants—and specific remedies or preparations were not recorded. The permissions to collect data regarding MPK restricted me from doing so (I elaborate a complete discussion of how I obtained permission to carry out this research later in the chapter).

Anthropologists continue to look at the various social and demographic factors that influence theoretical knowledge, as well as the relationships between economic activities and shared cultural knowledge. Past research on MPK has not emphasized a quantitative measurement of individual illness histories, but has instead relied on community-level availability and general usage of biomedicine versus non-Western medicine (Buenz 2005; Vandebroek et al. 2004). Based on the existing understandings of the dynamic relationships between experiences and specialized knowledge, I expected that there would be significant interactions linking MPK with economic activities and illness experiences.

Availability of different health services and therapeutic options influences an individual's use of services and knowledge of the respective medicine. Likewise, the socioeconomic profile of an individual influences the types of specialized cultural knowledge a person will have. Mexico is a vastly diverse nation, with many types of indigenous groups and a history of colonization that have helped contribute to a medically pluralistic society (Napolitano 2002). Health needs of communities are influenced by the types of livelihoods—the ways that people interact with their environments and make a living. It is because of these dynamic interrelationships between livelihood, health, and MPK that I explore the layers of local realities and expand to consider macro-processes implicit in the daily lives of people in the Reserve.

Model for the Study

My research for this book was carried out in a protected area among marginalized, self-identified "indigenous" communities in west central Mexico. The geopolitical location of the Reserve is significant not only because of the interestingly complex way of life in the communities studied, but also because of the explicit goals of sustainable development and conservation in the Reserve. There are various programming initiatives that stakeholders utilize to engage in sustainable development and conservation programming within the Reserve.

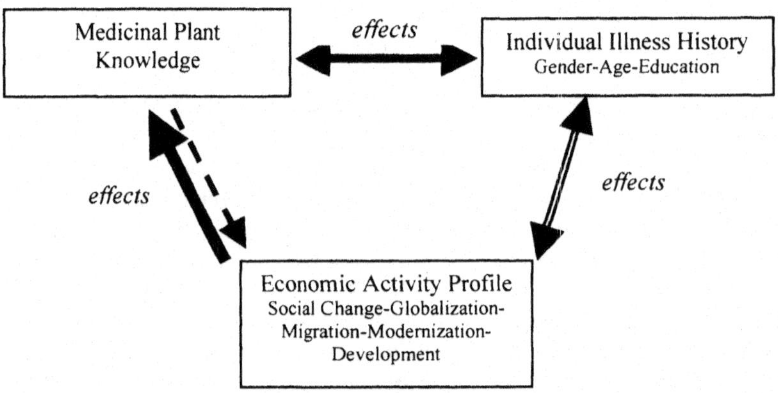

Figure 1.1. Conceptual model of the principle axes.

I carried out this research in the Reserve, which is a highly diverse environment in which ecological conservation and sustainable development have been focal activities (Benz et al. 1994; Gerritsen 2002). In the Reserve, as in other "developing" areas of Latin America, connections with the outside world are increasingly significant and influential at local levels. This book explores the connections between livelihoods, reported illness experiences, and differences in MPK in three communities within the Reserve. My research begins to explain variation between MPK and the economic profile and personal illness history. Figure 1.1 provides a rough visual illustration of the principle variables and their relationships. The goals of my research were to: identify economic activity profile, capture individual illness history, and test MPK, in order to better understand the various relationships (or *arrows*) that exist. MPK and individual illness history are likely to be interrelated since experience informs and contributes to the creation of knowledge and understanding. The bold arrow leading from "economic activity profile" to "medicinal plant knowledge" indicates that I expected a strong relationship of influence, whereas the dashed arrow leading in the reverse direction acknowledges that the influence in the reverse direction is limited or only seen in special cases (such as traditional healers).

The visual model in figure 1.1 helps to clarify the interrelatedness of all of the principle research variables, and the relative importance of each in addressing the research problem. Of course, these are not discrete variables, but overlapping, interacting, non-discrete aspects of life in the Reserve and Mexico. I expected that those individuals who work for waged labor and interact more with the regional markets, and who have had fewer illness experiences in which traditional medicine was used, would possess less MPK than other community members who work in the local forest area, and who have used traditional medicine more frequently in their illness experiences.

Objective 1a: Identify current MPK held by a representative sample of adult men and women in Ayotitlán, Telcruz, and Zacualpan.

Objective 1b: Determine significant patterns of variation in MPK.

Objective 1c: Record the life histories of self-described traditional healers or medicinal plant experts.

Objective 2a: Record the illness histories of research participants and their household members.

Objective 2b: Identify any relationships between the average illness histories in a community and the average MPK for each community.

Objective 3a: Explore the relationships between sustainable development, conservation, and MPK in the Sierra de Manantlán Biosphere Reserve.

Objective 3b: Identify variation in economic activity profile of households.

Hypothesis: Communities where the average household economic profile is characterized predominantly by non-farming activities will have less MPK on average than other study communities.

Figure 1.2. Study objectives and main hypothesis.

The overarching goal in my study is to describe cultural orientations to health and environment in the Reserve though a historical and political perspective, so I developed seven specific research objectives to rigorously answer how economic activity profile and illness experiences relate to MPK. My central hypothesis seeks to look for trends in the relationships between economic activities, illness experiences, and MPK in the Reserve.

Organization of the Book

There are three essential axes, or themes, in this book: socioeconomic aspects of indigenous livelihoods, health, and cultural knowledge (specifically, MPK). These themes are explored through an understanding of the historical, economic, and political complexity of indigenous people living in the Reserve. The objectives present a structured way of exploring the dimensions of livelihood, health, and cultural knowledge in the context of a protected area—a fact that constrains and influences the health experiences and livelihood strategies of peasants living in the Reserve. For these reasons, I work to articulate a detailed appreciation of the political and economic history of Mexico and the Reserve.

Identification of the fieldsite was an important component of my research. I needed to find an appropriate cultural area in which to study MPK and use of forest resources—a protected area was ideal. The remainder of this chapter will detail my process of finding my fieldsite and accomplishing the data collection. The methods used in this research are mixed—both quantitative and qualitative. The rationale behind the choice of these mixed methods and some comments on the specific procedures used are presented. In particular, I will discuss the relationship between myself as an investigator entering the community and the hurdles involved in that process.

I dedicate chapters 2 and 3 to introducing the cultural landscape and history of the Sierra of Manantlán and the current Biosphere Reserve. Within Mexico, there are various indigenous groups and geographical regions that cannot homogenized into a single short story. It is important to thoroughly understand the history of Nahua people in these mountains, their interactions with colonists, and current day interactions with each other and outside institutions. In the Reserve, the human presence dates back prior to the Nahua, to whom the indigenous people living in the Reserve claim ancestry (Laitner and Benz 1994). Though the Nahua language is no longer used on a daily basis in the communities, many of the forest resources are known commonly by their Nahua names. The focal sustainable development activities in the Reserve are centered on: the controlled extraction of forest resources; management of agricultural lands and livestock; and ecotourism (Gerritsen 2002; Jardel 1992). Included in this discussion is a description of the ways that people in the Reserve use natural resources, namely agriculture and animal husbandry. I also describe the ways that peasants in the Reserve make a living and how they relate to their forest resources. The primary goals in this chapter are: to identify the economic activity profiles prevalent in my study sample, to describe similarities and differences within and between sample communities, and the representativeness of my sample. It is important that my study sample be representative of their communities so that my model has applicability to understanding trends of sustainable development and conservation in the Reserve.

Chapter 2 provides a brief history of the people in the Reserve, and a description of the society and culture of the study communities. The local cultural traditions are important to understanding the identities of people living in the Reserve. The prevalence of Catholicism, resulting from a history of colonization and religious syncretism, has shaped modern identities in the Reserve. Though there is not a lot of detail known about the region prior to colonization, we do know some about the indigenous roots. I also offer this discussion in light of larger trends and historical events in Mexico. Later in chapter 3, I discuss the role of conservation and sustainable development in Mexico, and specifically on a biosphere reserve. The relevance of sustainable development and conservation goals is central to understanding MPK in the Reserve because of the lived realities created by being on such a protected area.

In chapter 4, I delve into the theoretical and methodological questions of cultural knowledge and economic activities related to livelihoods. The freelisting methodology I used to help measure cultural knowledge is important because of the simplicity of the device and its utility in anthropological investigations. Owing to its versatility, and the inherent simplicity of the freelisting technique, it is good for measuring intracultural variation in knowledge (Quinlan 2005).

Returning to the theme of natural resource management in a protected area, chapter 5 describes the ways that the Reserve is constituted and governed. I give some examples of the different ways that conservation and sustainable development programming is being carried out in the Reserve in chapter 6. I discuss these examples in terms of the sociopolitical dynamics of decision-making and accountability.

Chapters 7 and 8 switch gears, as I focus on the third and final axis of this book: medical pluralism and health profile in Mexico and the Reserve, specifically. I look at the general health profile and health priorities in Mexico because they shape the policies and programming of services available in the Reserve. Furthermore, the health needs of the communities in the Reserve reflect the broader trends in marginalized communities in Mexico, so it is appropriate to supplement the primary data collected for this study with national level, secondary, data. I look at mothers and children as indicators of the health in a community and discuss the ethnomedical pluralism present in contemporary Mexico. I also discuss the prevalence of medical pluralism in the Reserve where individuals choose different treatment pathways.

Continuing the ethnographic description of ethnomedical systems in the Reserve, I present the results of my research of traditional medicine and traditional healers in chapter 8. Using the in-depth interviews I conducted with key informants who are considered experts in traditional medicine in their communities, I demonstrate the reasons for the persistence of the perceived value of traditional medicine and medicinal plants. I also present the interviews conducted with biomedical practitioners in the study communities, providing a context for understanding medical plurality in the Reserve.

Evaluating the study hypothesis is important because it illuminates some relationships between the three central axes—health, livelihood, and conservation—that contradict common assumptions about traditional knowledge and modernization. For example, I found that those individuals who have used traditional medicine less frequently in their own illness experiences are not more likely to have less MPK. There is not a clear, linear relationship between the amount of MPK community members have, on average, and other socioeconomic indicators. I explain how my results compare to prior studies regarding MPK in the Reserve, and in other societies. Since there is no linear model associating the loss of traditional knowledge with increasing modernity, the contribution of my book is first and foremost as a case-specific ethnography.

Finally, in chapter 9 I unpack the significance of my research for the community members in the Reserve through the example of the Traditional Medicine House in Ayotitlán. I use this example to illustrate the potential value of thinking about development and conservation projects as sets of trade-offs and hard choices, rather than seeking the ever elusive "win-win" strategy.

Fieldwork in the Sierra of Manantlán

The remainder of this chapter provides a detailed account of my fieldwork activities during the data collection period. I conducted the research for this study in the Reserve from January 2007 through May 2008. As I previously indicated, my research for this book aimed to: identify economic activity profile, capture individual illness history, and test MPK in the Reserve. My fieldwork had four stages: (1) preliminary preparations and orientations to the Reserve, (2) collection of raw data, (3) construction of study variables and analysis of model, and (4) dissemination of the study findings. The first stage of the fieldwork was essential to constructing the research instruments and achieving permissions to carry out the data collection. Next, I collected raw data on economic and social activities and personal illness history, and administered the test of MPK. The third phase of my research was the analysis of data and construction of indices for the variables. Here, I focus on the methodologies during my first two phases of fieldwork; chapter 8 deals extensively with the construction of the research indices and looking for trends. In the first phase of research, I used three primary modes of data collection: (1) household surveys; (2) individual tests; and (3) life history interviews of key informants.

Preliminary Preparation and Orientations to the Reserve

Once I had identified my research topic of MPK and its relationship to so-called "development" and "modernization," I began to look for a good field site. I had read Bruce Benz's and colleagues' work on inventorying forest resources in the Sierra of Manantlán. In one critical paper (Benz et al. 2000), they attempted to account for loss of indigenous plant knowledge in relationship to modernization and other socioeconomic differences within the Reserve. Their research, however, did not lead to conclusive results-perhaps partially a result of using secondary and aggregate data instead of looking for variation within communities at the household and individual level (Benz et al. 2000). I realized that my research could build directly on the studies that they had already done in the Reserve, so I sent an e-mail to Dr. Bruce Benz. He agreed about the potential scholarly contributions of my research, and via electronic communication directed me to the town of Autlán de Navarro, where my fieldwork story begins.

My fieldwork began in January 2007 when I began to simply get to know the community of Sierra of Manantlán—its organizations, institutions, communities, and people. I began by simply being present and available to the community members to answer questions or get acquainted. I focused on *listening*, not asking too many questions—introducing myself and my interests when asked (which was quite often). I was simultaneously absorbing and being absorbed. Doing so enabled me to avoid wandering in and asking questions that would have been unprecedented and, in all likelihood, answered dishonestly. In fact, one of the first things people would repeatedly tell me was "Pues, es que siempre vienen de alla. Llevan las plantas y se van," (Well, foreigners always come [from the USA and Canada]. They take plant [samples] and they leave). The notion of ongoing cultural and material imperialism is palpable in Ayotitlán (one study community), and to lesser degrees in the other study communities.

The process of obtaining approval to work in the Sierra of Manantlán communities began in January 2007 when I contacted the Management of the Biosphere Reserve and the University of Guadalajara South Coast Campus (CUCSUR, from the Spanish *Centro Universitario de la Costa Sur*). At CUCSUR, there is a core interdisciplinary team of researchers at the Manantlán Institute of Ecology and Biodiversity (IMECBIO). The IMECBIO (part of CUCSUR), located in the urban center of Autlán, Jalisco, is the interdisciplinary research team that measures and leads assessments of the interactions between humans, environment, and organizational programming throughout the Reserve. Preliminary participant observation and integration into the community were beneficial in two primary ways. First, I developed relationships of recognition, familiarity, and trust in the community. I received an official letter of cooperation from the Director of the IMECBIO and developed a working relationship with the Management of the Reserve.[1] Through these relationships with relevant institutions, I used appropriate channels to contact the Ayotitlán Community Council and presented my research prospectus to them at their May 2007 meeting (I eventually received permission to carry out the study in Ayotitlán in early November 2007). Ayotitlán and Telcruz are within the same *ejido* (governing and decision-making unit). I also made presentations to the Zacualpan Commission of Communal Goods (elected leadership) in May 2007.

Second, this period of preliminary work informed the development of the research problem. Spending time in the community illuminated the daily reality of domestic migration, the importance of cultural patrimony, and the environmental concerns that swing in the balance. It became clear early on that community members are actively negotiating these concerns and (re)prioritizing. I presented a Spanish version of my research prospectus to both the Director of the Reserve and the Director of the Institute on Ecology and Biodiversity at CUCSUR, as well as a number of other instructors and researchers at CUCSUR. During this process, I had lengthy discussions with local academic stakeholders in the CUCSUR regarding my research propositions. These extended discussions were particularly helpful as I selected my sample communities (Zacualpan, Telcruz, and Ayotitlán) and devised a strategy for collecting the data necessary to answer my research question. These three communities were selected because of their variation in proximity to regional market centers (Colima, Manzanillo, and Autlán) and their varying degrees of modernity and marginalization. They were recommended to me by these local researchers who are very familiar with the various communities in the Reserve, and I was cautioned to have a back-up community for Ayotitlán which is considered very difficult for carrying out research. Ayotitlán is the most marginalized and most "indigenous," Zacualpan one of the most modern, peri-urban of the communities in the Reserve, and Telcruz was included as a gradient in between Ayotitlán and Zacualpan. By September 2007, I received final approval from the Institutional Review Board of my home university, Case Western Reserve, for all of my instruments (questionnaires and semi-structured interviews), my processes and procedures for obtaining informed consent from community members, and my research announcements to the communities.

I dedicated months to the process of obtaining prior informed consent to work in the three communities. This process was in itself very informative of the social and political dynamics within each community. Whereas it was relatively easy to obtain consent to work in Zacualpan, the Ayotitlán and Telcruz communities (both within the same ejido) were much more skeptical of my proposed research. I maintained field notes throughout this process regarding my interactions and discussions with all stakeholders.

Throughout the process of becoming engaged in the communities and learning about local needs, I became involved with a wide variety of secondary activities. I call them *secondary* because they are not directly related to answering the research question—rather, they are applied anthropological activities. I decided that I would work to ensure that my presence in the Sierra of Manantlán would have some positive impacts. One of the largest projects I have begun involves a women's collective group in a neighboring community of Cuzalapa (about 20 km from Ayotitlán). I used much of my "down time" to help create a website for them (www.cafecuzalapa.com) and to recreate some of their promotional materials. Performing these services gave me the opportunity to interact with community members in a positive way while not directly impacting my sample or research sites. I learned from these women in informal participant observation how daily life in this region varies across households.

Sampling Strategy

I used a stratified sample to test across generations in the cross-sectional part of my investigation, and I used purposeful sampling for the qualitative data collection. Data for the household variables were collected from the main study sample, which was divided into three cohorts to obtain a representative sample (around 3 percent) of adult men and women in the communities of Ayotitlán, Telcruz, and Zacualpan (shown in figure 1.3). In Ayotitlán, only three men between the ages of thirty-six to sixty-five were able to participate, which could very likely be a result of out-migration for labor in that demographic category (labor migration is discussed briefly in chapter 2).

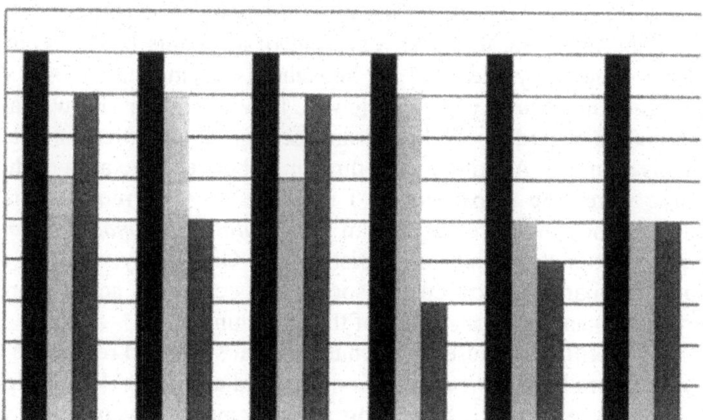

	Women 18-35	Men 18-35	Women 36-65	Men 36-65	Women 66 +	Men 66 +
Zacualpan	9	9	9	9	9	9
Telcruz	6	8	6	8	5	5
Ayotitlán	8	5	8	3	4	5

Figure 1.3. Distribution of study sample by age and gender.

In October 2007, I began conducting interviews intensively in Zacualpan. At that point, I still did not have permission to work in Ayotitlán (it was obtained in early November). The backup community was going to be Cuzalapa if I was not able to obtain permission to research in Ayotitlán. Cuzalapa had been used in previous research studies conducted by IMECBIO and is located about one hour from Ayotitlán in a neighboring valley. Cuzalapa would not have been as ideal because the community is not as marginalized as Ayotitlán. Harvesting of maize begins in late November, so I was pressured to start rapid collection of the household survey data.

Data collection began in full swing by November 2007, and sampling method was an important concern. Participants were not selected at random from the census. Because I wanted to have a specific number of men and women in different age categories, and because of the strenuous process I underwent to simply enter the communities, I worked with my field assistants (two in Zacualpan and three each in Telcruz and Ayotitlán) to sample using a snowball technique. In Zacualpan, we began by going house to house, and later targeted households where one of the informants was able to estimate that someone in the household was of the desired sex and age to fill the sampling categories.

I used a purposeful sampling technique to obtain a stratified sample representing different ages and genders in the three study communities, and I also used purposeful sampling for the qualitative data collection. Data for the household variables were collected from the main study sample, which was divided into three cohorts to obtain a sample of adult men and women in the communities of Ayotitlán, Telcruz, and Zacualpan (shown in table 2.2). I sampled nearly 14 percent of households in Zacualpan, 15 percent in Telcruz, and 23 percent in Ayotitlán, including just over 3 percent of all adults in each community in my sample (since only one adult per household was included). Only one individual from each household was interviewed, however he or she was asked to report information regarding livelihood and economic activities for the entire household as a unit. In Ayotitlán, only three men between the ages of thirty-six to sixty-five were able to participate. I used the most current national census data (2005) at the time of research from the *Instituto Nacional de Estadística y Geografía* (National Institute of Statistics and Geography; INEGI) to complete a rough comparison between the social and economic activity profile of study participants and the marginality of the community.

In Ayotitlán, because of internal political strife and resistance to presence of any outsiders by many, we relied on my field assistants' knowledge of individuals who were likely to be willing to participate. Nevertheless, we found that some elected not to participate and almost all were hesitant about the study's objectives and outputs. Some of the confusion had to do with the timing—in October 2007 the national agricultural and animal husbandry census had been undertaken. This census caused some people to be suspicious of me and my research assistants when we came in late November asking many similar questions regarding who lives in the household, level of education, agriculture and animal husbandry activities.

I also conducted semi-structured life history interviews with key informants who were selected purposefully, based on their location of residence and/or employment, the nature of their employment or reputation as medicinal experts, and agreement to participate. I included the surrounding community of Cuzalapa, where I found a traditional healer who was interested in completing the semi-structured interview with me.

Quantitative Data Collection Methods

The MPK was an individual level variable, but the livelihood information was reported at the household level. The presentation of my research in this book is organized around the set of objectives and central hypotheses provided in figure 1.2. I employed both quantitative and qualitative methods to appropriately collect the information necessary. I used freelisting, semi-structured interviews, household surveys, and participant observation. Utilizing these techniques in conjunction is the most effective way to identify emic domains (Quinlan 2005).

The primary surveys were conducted in early December 2007 with the assistance of trained research assistants from each of the respective communities. In the end, this was the most feasible way to carry out the surveys—because of the desired sampling frame, it was not possible to conduct random sampling. The study sample is described in chapter 3.

The quantitative data were collected using a ten-page formal questionnaire that was administered to a total of 125 participants. The last page of this questionnaire was a freelisting of all medicinal plants the participant could name, followed by the question of whether he or she had ever used the particular plant. The rest of the survey was organized into six sections. The first section in the demography portion of the survey collects personal and family information that can identify and charactize the entire household. I recorded for all members of the household their age, relationship to the primary participant, level of education, and profession. I have maintained a master list of all individuals with the anticipation of a future follow-up study with some of the younger generations from the same households to capture longitudinal data on socioeconomic dynamics and MPK. It would be very beneficial to return to these same communities and measure the same variables again in five or ten years. I could even (ideally) measure the MPK and socioeconomic status of some of the children of those who participated in this first study. A longitudinal study design will illuminate the processes related to development, conservation, and diffusion of cultural knowledge in the Reserve.

The second section captured detailed demographic information regarding only the primary informant—in particular, the total years of education (at what age the participant first attended school and for how many consecutive years). Redundancy in the questions captured various measures of the same variable, helping to reduce informant inaccuracy or other sources of error in the surveying. Other questions included where the informant was born and had lived, visited, or worked. Regional market towns of importance are Autlán, Manzanillo, and Minatitlán—although there are different degrees of accessibility to, and interaction with, these three different market towns because of the distinct locations of the three study communities.

The third section dealt with household characteristics, from the size and materials from which the house was constructed, to the small, yard-garden (*traspatio*) and presence of pigs, chickens, ducks, or other yard animals for household consumption. The size of the house, yard, and the types of vegetables and fruits grown in the yard (if any) are characteristics essential to and indicative of socioeconomic status because they are part of wealth and quality of life.

The fourth section of the household survey was on agriculture and animal husbandry—specifically, cattle ranching. Here I used an exhaustive surveying technique to ask about all fields and cattle held, used, bought, sold, traded, or born in the last year. I asked about the quality of the crops, health of the cattle, and how the crops were grown. Was a tractor used to plough? Did they spray insecticides or fertilizers? Was the corn this year better, worse, or the same as the previous year? How many seasons a year do they plant? The questions covered all aspects of farming and cattle raising—mimicking other exhaustive agriculture and cattle surveys used by anthropologists.

Economic activity was measured through an exhaustive structured interview regarding all economic activities within the household for the two weeks prior to interview: consumption, purchases, incomes, sales, and exchanges (Godoy et al. 1998; Reyes-Garcia 2001). The data were collected at the household level because that is the most appropriate manner to capture the socioeconomic status of an individual. I asked about all types of economic transactions over the last two weeks or one month. In this section on economic activities, I asked about waged labor, items bought, sold, traded, debts and loans acquired or paid, and remittances. The data I collected regarding the livelihood of the individual and household help to understand the socioeconomic status and interactions with different components of markets. Integration into the modern biomedical health care system occurs through various aspects of daily life for individuals in marginalized communities. Just as the type of work an individual does to make a living influences his or her health needs, the exposure to aspects of modernity also influence an individual's specialized cultural knowledge. Livelihood and socioeconomic status are frequently found to be interrelated with health status, particularly in marginalized communities.

Finally, I asked a series of questions regarding the informant's personal health history. I asked how many times in the last week they had drunk alcohol (either beer or other spirits) or smoked tobacco cigarettes to help approximate one possible aspect of the current health status of the participant. Not only do these questions help provide clues as to the overall health status of an individual, but also reflect exposure to public health campaigns aimed at reducing the use of alcohol and tobacco. Public health campaigns in Mexico are often approached through television and radio ads or community health fairs (in addition to the school curriculum directed at youth). I also asked about any sickness he or she had had in the last six months, including all treatment pathways and their associated financial and time costs. Structured interviews recorded illness episodes experienced by the participant for two time frames: the six-month period preceding the interview, and the entire life of the informant. For each illness episode, the informant was asked: (1) what therapeutic pathways were sought; (2) the costs of the therapy; and (3) whether the results were positive. The informant was also asked (4) what he or she did next, until indicating that no other action was taken. All illness episodes in the six-month period preceding the interview were recorded, but for the entire lifespan of the individual I only asked about major illness episodes. Again, for the major illness episodes I asked what treatment pathways were taken, the costs and outcomes of these treatments, and so on, until he or she indicated that no further action was taken. The reason for only asking about major illness episodes in the lifetime is to avoid recall error; however, for the six-month period preceding the interview informants were expected to be able to recall with confidence all illness episodes. Lastly, I asked them to rank their own health now as compared to one year ago—is it better, worse, or about the same?

The census inventories I used are modeled after classic household census surveys used by other anthropologists who are interested in evaluating the relationships between livelihood, health profile, and MPK (Godoy 2001; Reyes-Garcia 2001). Freelisting was used to quantitatively measure the dependent variable: the number of medicinal plants an individual can list. Freelisting allowed me to assess the quantity of MPK held by individuals, which I used as a proxy measure of individual cultural knowledge. Freelists were collected from the same 125 informants who participated in the socioeconomic census survey of their respective households. Some of the freelists were conducted early on in the forty-five to sixty minute interview, but most were completed as the last segment. I began to move the freelist test to an earlier part of the survey because I noticed that informants could lose interest after about thirty minutes of interview. I did not lead the informants with prompts such as "Can you think of any plants for *X*?" Doing so would have created inaccuracy in the results because it would limit the domain of knowledge being tested. Instead, I asked for any medicinal plants for any illness, *cualquier planta que se puede usar para curar* (any plant you can use to cure). I did not use successive freelist probing, but simply repeated the initial task at least three times to encourage the informant "Can you think of any other medicinal plants?"

All plant names were coded into numbers immediately (with the original lists being stored securely with the other master lists from the study) in order to respect the integrity of indigenous knowledge of these communities and according to our agreement permitting my research study. Abiding with that agreement, no plant names are mentioned in this book. It is not necessary for the names of plants to be listed to understand whether or not there are differences in plant knowledge in the Reserve. Future applications of the results presented here may seek ways to actively promote conservation of local MPK.

Qualitative Data Collection Methods

I administered a semi-structured survey to individuals who self-identified as working with either Western biomedicine or traditional medicine in the communities. They included nurses, doctors, traditional healers, and knowledgeable neighbors. The objective was to capture a partial life history of these individuals as related to health, medicine, forest resources, traditions, and culture. I specifically asked for their personal and professional experiences and opinions. I used a semi-structured interview to elicit the major life events and factors contributing to the individual's path to becoming a self-identified traditional healers or expert in medicinal plants. The questions I asked dealt with successes, areas of expertise, who they have learned from, and perspectives on the use of medicinal plants, which were all discussed in an unfolding of the individual's life history.

Participant observation was carried out in the community and structured field notes were recorded. I kept field notes on a regular basis beginning in January 2007 and continuing through May 2008. Field notes were primarily based on my observations, interactions, and experiences living in the Reserve. The content of the field notes included: the nature of forest usage, cultural traditions, and the local systems of governance. My field notes provided the contextual "reality check" of the quantitative survey results, and thereby help tell the story of the Reserve and the individuals who live and work in the different communities there. They inform my interpretation of the political and social processes that I participated in while seeking consent to work in the communities. The field notes also capture the depth of shared experiences in the various communities; the relationships I formed with individuals and certain families are the center point of my relationships outward in the communities. I collected field notes like a field journal, and I took care to always note: (1) date, (2) time of start and duration, (3) place, (4) people present and involved, and (5) primary activities and outcomes. The field notes recorded meetings and conversations and experiences I had throughout the phases of data collection. They were not always recorded daily, which was not a weakness but rather facilitated focusing on the truly essential moments of my fieldwork and meaningful introspection. The following principal qualitative categories emerged:

- Social relations
    - Economic and household issues
    - Personal relationships of individuals associated with, or living in, the Reserve
- Sociopolitical processes in the Reserve
    - Problems within specific communities
    - Programs and initiatives
- Healthcare provided in the Reserve
    - Use of traditional medicine
        - Personal experiences with traditional medicine
    - Use of modern biomedicine

- Personal experiences with modern biomedicine.

These categories are not discrete but are overlapping. The primary objective of the field notes I collected was to inform a descriptive ethnography of the people, places, and processes involved in the Reserve—specifically related to medical systems, but not strictly. These field notes are already able to explain much of the relationships between the macro- and micro-sociopolitical processes.

Quality of the Data

Measuring Knowledge with Freelisting

I considered using different tests for measuring MPK in my study. I had to change the frame elicitation task to a freelist. It was too hard to get people to recognize a photograph of a plant, and I did not have permission to collect herbarium voucher specimens. Initially, I was hesitant about the efficacy of using the freelist task which a professor in the botany department at CUCSUR recommended. In order to protect intellectual property rights of individuals and communities, and because I did not have permission to collect plants within the Reserve (I never attempted to gain this permission), it was clear I would have to work with either knowledge tests using photos of plants, dried plant samples acquired through purchase, or use a freelisting technique. In the end, I opted for the freelisting technique because (1) I cannot use the names of any of the plants, so there would have been a lot of time and energy lost acquiring photos or dry samples; and (2) the qualities of the plants listed are not essential to my research problem. I wanted to know the quantity of plants an individual could list—how much MPK a person has. I could use the list length as a proxy measure of MPK.

Methods for Economic and Health Information

In indigenous populations in Mexico, particularly those of Manantlán, it is socially undesirable to complain about past illnesses. For this reason, it seems that most individuals underreported their illness episodes in both the previous six-month and lifetime periods. It is likewise socially undesirable to "brag" or be perceived as having more wealth than others; the economic data also reflect varying reports of remittances and government stipends. I can predict that there were both more illness episodes and more remittances and stipends than reported during the interviews.

Recall error of illness experiences is likely to be a significant factor in the results. Adults were asked about all of their own illness episodes over the last six months, but this is an inherently difficult task. Adults are very likely to omit a common cold or cough—even if it was something that kept them from their normal activities for a couple of days and caused them to seek some sort of medicinal remedy. It might have been more effective to ask about illness episodes of any members of the household over the last six months. Children not only get sick often, but the severity of even a common cold can be much greater because of their fragile immune systems. Adults are preoccupied more with the health of their children than with their own, in many instances. Nonetheless, because of the costs of accessing health care (travel and expense of medicines) many adults were prone to remember illness episodes from the prior six months.

Omission or non-response is another important source of error. Some community members are suspicious of outsiders due to past experiences. Though all data were collected in teams and pairs including young adults from the communities being surveyed, there is still a degree of omitted or missing data. Furthermore, it could be reasonably speculated that some of the information provided in the household characteristics and economic question sets was underestimated or simply not reported. In Zacualpan, there was overall less omission or non-response to survey questions; Ayotitlán demonstrated the most inconsistency and non-response to survey questions.

Errors in Coding

I performed all coding and data entry myself during the months January to July 2008. Errors in the codes of the data presented here are most likely to be found in two forms: (1) incorrect spelling on survey questionnaire leading to miscoding, or (2) typos and errors in data entry. I tried to minimize these errors by careful coding, consulting primary informants to verify spellings of plants, and reconciling the original survey questionnaire when in doubt.

All coding was done manually, especially since many of the same plant names have different spellings. I kept a master list of names and all members of the same household have the same code sequence. I created a master coding manual during this process, first entering all data into Access Database, then into Excel workbooks where they were cleaned and coding finalized, and then into SPSS (a statistical analysis software package, acronym originally referred to *Statistical Package for the Social Sciences*).

Reciprocating with the Study Communities

Establishing Permission to Work in the Field Sites

Outside researchers and scientists, including those from the IMECBIO, are expected to go through the process of obtaining prior consent from the community to carry out research. This component of the research and programming process speaks to issues of legitimacy of the process itself (Brechin et al. 2002). This process is particularly important in Ayotitlán because in many other communities, such as Zacualpan, the process is less critical and more of a formality. In Ayotitlán, the Council of Elders speaks up for a community that is otherwise largely quiet during assemblies and meetings. A few of the members of the

Council of Elders—seeing it as their moral responsibility—do question the goals of outside researchers. In this process, they ask what the primary activities will be and how the community is expected to benefit from the research activities. Additionally, they continue to demand copies of all research work and publications resulting from research carried out within Ayotitlán. In asserting the proper, legitimate process for interacting with outsiders, the community of Ayotitlán demonstrates strong organizational and institutional capacities.

Informing the Communities About the Research

At the completion of data collection (which lasted about sixteen months), I reported initial findings to community representatives. The process allowed me to share insights regarding possible ways to conserve certain aspects of their cultural heritage and promote health for themselves and their environment. I presented digital and print copies of my doctoral dissertation to each of the study communities in March 2010. Giving copies of my work to the communities is not really a final step, but an acknowledgment of gratitude, respect, and a desire for continued interaction. Perhaps most importantly, doing so was part of the initial agreement I made with the communities when I was granted permission to carry out the research.

Benefits of the Research to the Communities

Working in marginalized communities is difficult because of the researcher's often unrealistic desire to leave tangible benefits for the community. My acceptance into community members' homes, community meetings, and events, and the information shared with me is larger than the reciprocation I have been able to give. However, I do hope that there have already been some benefits, and that through future collaborations and work there will be more to stem from my research/activist relationship with these communities. My insistence on following proper protocols for establishing permission to work in the communities seems to have brought attention to other research being done in some communities that does not adhere to this protocol. When the community members saw that I was willing to endure months of seeking permission to work in the community, they intensified their challenge to other researchers and demanded long-overdue copies of past research studies carried out in their communities. Second, I think that the questions I asked helped re-open old topics that are of interest, in varying degrees, to community members.

I began working with community development projects and ecotourism projects in the Reserve. I have created new brochures and advertisements for the women's group (Café de Cuzalapa) whose members produce shade-grown coffee and various arts and crafts. I helped to create, and continue to maintain, a new website for them. These are activities I have taken up on the side that are enriching my experiences with the people and groups of the Sierra of

Manantlán. Working with groups like this women's group has given me access to different aspects of the political structures that are operating within the different communities.

Involving Community Members in the Research

I worked with eight different field assistants from across the three research sites. They gained training and fieldwork experience (along with a good pay rate). I gained more reliable results (because these university students are members of the communities they worked in) and help in completing all of the surveys within a six-week timeframe between the harvest and Christmas holiday season. There were various advantages and benefits from working with field assistants in each community; the students gained valuable research training and experience, and were also paid a good wage for their time, which together encourage the continued higher education of these students.

This is an anthropological study that centers on individuals and community groups as primary actors. The focus of analysis takes into account broader, macro-level processes that influence the local-level realities in the Reserve (such as bureaucratic processes and policies). Though the study communities are of indigenous, Nahua origin and descent, they are not isolated from national Mexican politics nor from regional and global economic markets (Olson and Gerritsen 2011). Though difficult, it is essential to acknowledge the implications of broader market forces and political ideologies on the individuals and families who are stewards of the Reserve (Gerritsen 2002). Conservation is thus viewed as a social act which is embedded in other anthropological dynamics (Brechin et al. 2002).

Notes

1. References to the Management of the Reserve in this book are directly to the field representatives who are essentially the "federal forest rangers" working directly in the communities to manage forest resources. These field representatives come from disparate parts of Mexico and are rarely from the local area (at the time of research only one of the field representatives was from the Reserve and had his family still living on the Reserve).

Chapter 2
Sierra of Manantlán Biosphere Reserve

The ethnography begins by describing the place: the Sierra of Manantlán region, which has been inhabited since prior to the period of contact and colonization. I begin by characterizing the geographical area of the Sierra of Manantlán, as I turn to a brief overview of some of the key historical events in the region. I also describe some important social and demographic attributes of the indigenous peoples living in the region. In the second half of this chapter, I describe the demographic, economic, social, and cultural dimensions of the three study communities. I describe the cultural traditions and customs of Ayotitlán, Telcruz, and Zacualpan. Through these descriptions you can gain a sense of the range of identities of individuals living in the different study communities.

Geographic Area

The Sierra of Manantlán is located near the southwest Pacific coast in Mexico, straddling the states of Jalisco and Colima. The name has been used to refer to the indigenous communities of Nahua origins who continue to live throughout the region.[1] Much of this territory is now a federally mandated Sierra de Manantlán Biosphere Reserve (in Spanish *Reserva Biosfera de la Sierra de Manantlán*), which is also part of the UNESCO Man and the Biosphere program. Mountainous, deciduous evergreen and tropical deciduous forest types account for the vast majority of the land area (see table 2.1). Permanent agricultural land was only a small fraction of the total area (2.7 percent in 1993). In 1978, the genetic ancestor of corn (*Zea perennis*) was found in the Sierra of Manantlán. *Z. perennis* is more commonly referred to by its Nahua name *teocintle* and brought the Sierra of Manantlán to international fame for its botanical and biological gems (Benz et al. 1994; Jardel 1992). Teocintle is to corn what wild wheat grass is to wheat—a cornerstone in the historical development of humankind in the Americas (Von Hagen and Beltrán 1961).

The region is most easily reached by the Guadalajara, Jalisco international airport, from which a three- to four-hour bus ride leads to Autlán de Navarro, the common entry point to the Reserve. The Reserve can also be reached through the urban center of Colima, in the state of Colima (see figure 2.1). There is significant variation in the degree and type of interactions members of the primary study communities have with surrounding urban areas, particularly Autlán, Minatitlan, Manzanillo, and Colima, shown in figure 2.2. This region is also characterized by the presence of the Jalisco-Colima volcanoes, one of which is still active and is a potential threat to many of the communities in the Reserve.

Figure 2.1. National map of Mexico and the Reserve. Copyright E. A. Olson.

Biodiversity in the Sierra of Manantlán Biosphere Reserve

The Reserve covers an extensive area (139,500 hectares) representing the largest and most important protected area in western Mexico (INEGI 2005). The diversity of habitats is also impressive, with altitudes ranging from 200 to 2,860 meters above sea level. There are nine primary types of vegetation and the Reserve is located very close to the Pacific Ocean, creating a humid, tropical environment in the western and southern parts of the reserve (Graf et al. 2003; Carabias Lillo et al. 2000). Pine and mixed pine forests at higher altitudes have dry, cooler climates. There are six general types of forest and three classifications of other vegetation within the Reserve, however more discerning classifications designate more specific types of forests and land vegetation within these primary types (Carabias Lillo et al. 2000). The three most prevalent types of vegetation are forests (as shown in table 2.1): deciduous evergreen; evergreen-mesophytic mountainous; and tropical deciduous.

The three study communities I worked in (Zacualpan, Telcruz, and Ayotitlán) are situated in predominately tropical deciduous forests, which is the second most common type of vegetation in the Reserve (Vázquez and Cuevas 1995). The plant species variety in each of the study communities is comparable and there is not a significant amount of variation in the types of common plants available in any of the communities. The fields of community members of Zacualpan are almost entirely in tropical deciduous forests, while the town where most reside is centralized and peri-urban. The fields, and community, of Ayotitlán are located almost entirely in tropical deciduous forests, as well. Telcruz demonstrates a slight amount of variation because in and around Telcruz there is more secondary vegetation and pastureland. However, the area is still heavily classified as tropical deciduous forests and according to plant biologists at the IMECBIO there would not be sufficient amount of variation in the types of plants and vegetation across the three communities, (Santana Michel and Cuevas Guzmán 2007).

Table 2.1. Surface area covered by the principal land and forest classifications in the Reserve in 1993.

Type of Vegetation	*Surface Area (%)*
Pine-Evergreen Forest	15.6
Evergreen-Mesophytic Mountainous Forest	18.5
Mesophytic Mountainous Forest	1.5
Tropical Subdeciduous Forest	2.1
Deciduous Evergreen Forest	19.6
Tropical Deciduous Forest	18.4
Thicket-Pasture Land	11.3
Pastureland and Sloping Agricultural Land	10.3
Permanent Agricultural Land (irrigated and seasonal)	2.7
TOTAL	100.0

Source: Carabias, Julia, Javier de la Maza , and Rosaura Cadena. 2003. *Capacidades necesarias para el manejo de areas protegidas: America Latina y el Caribe.* Arlington, VA: The Nature Conservancy, page 55.

A maintained gravel road that runs from Cuautitlán, the municipal seat, provides access to Ayotitlán and Telcruz. Daily buses pass by Ayotitlán on the main road and connect to the highway in Cuautitlán, which is reached via dirt road in about one hour by truck. From the municipal seat, more frequent buses run to the regional centers of Autlán and El Grullo (1.5 and 2.5 hours away, respectively). The same bus that passes through Ayotitlán, about one hour further down the dirt road, accesses Telcruz. It is easier to access Telcruz from the small town of Minatitlán. Telcruz is nestled deep in the Reserve, and has a quieter character owing largely to its remote location. This same dirt road can be continued about another hour to reach the highway, from which Zacualpan sits another twenty minutes further northeast. However, the dirt road is not always in a drivable condition—often requiring 4x4 to pass—and the bus does not continue past Telcruz. Instead, Zacualpan is most easily reached from the city Manzanillo, or Minatitlán, using the buses that operate frequently to and from Colima. From Colima, a second-class bus makes the twenty-minute trip twice an hour to and from Zacualpan.

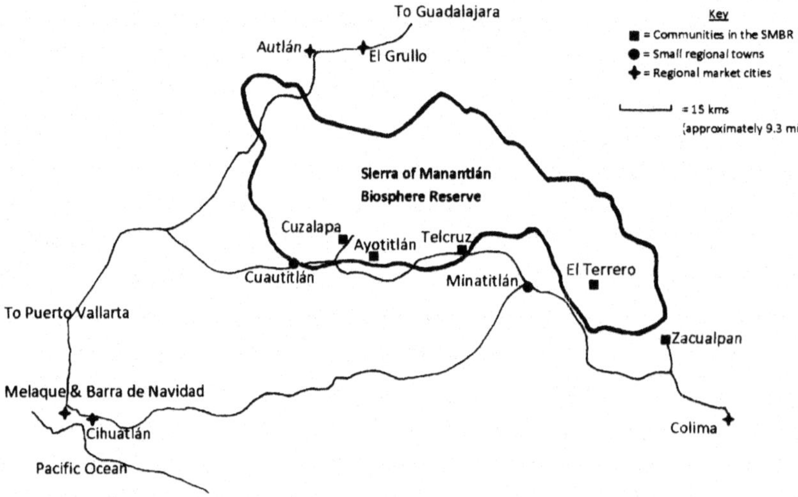

Figure 2.2. Regional map of the Reserve and urban centers of influence. Copyright E. A. Olson.

The communities of Zacualpan, Telcruz, and Ayotitlán in the Reserve have been reported to have statistically significant socioeconomic variation (Benz et al. 1994; Gerritsen 2002). Ayotitlán, the most marginal of the three study sites, is a community of around 700 individuals, while Telcruz has about 1,200 inhabitants and Zacualpan just over 1,700 (INEGI 2005). Living in the humid lowland forests of the Reserve, these communities rely on mixed subsistence strategies. Many families rely on a mixture of low intensity agriculture, waged labor in the nearby sugar, cattle, or coastal tourism industries, and subsistence horticulture. In all three of my study communities there is access to modern services such as electricity, running water, and sewers. These services are not always reliable and there is limited telephone (and no mobile phone) access in Ayotitlán and Telcruz. Zacualpan is the most integrated with regional and other external markets.

Early Presence of Nahua

In the pre-colonial period, the region of the Sierra of Manantlán encompassed five provinces: Amula, Milpa, Tepetitango, Cihuatlan, and Espuchimilco (IMECBIO 2000). Across these provinces, however, there was and continues to be a great deal of diversity. This heterogeneity comes in environmental, climatic, and socioeconomic dimensions. Among these, the province of Milpa (the Aútlan–El Grullo valley) had the highest degree of urbanization and intensive agriculture and ranching. In contrast, the province of Amula had the lowest levels of urbanization (Laitner and Benz 1994).

Milpa (shifting maize cultivation) is a term that refers to a specific type of agriculture predominant in the Reserve; it is a Nahua term that means "to the field" (Mann 2005) and is a system of growing maize with squash and beans in the same field. The corn plant grows quicker and its stalk serves as a pole for the bean and squash vine to climb. Fields are located anywhere from two to fifteen miles from the homes of peasant farmers. In forested areas, the trees are cut and then felled the next year, which allows the ash to mix with the topsoil and fertilize the field (Von Hagen and Beltrán 1961). The process of milpa does not require high levels of technological intervention since the rotational and mixed crop strategies permit the farmers to avoid using costly chemical fertilizers or pesticides. The system typically requires allowing the field to go to fallow for eight years after two years of cultivation, thus it is a sustainable strategy for small, local producers but would not be a viable solution at a larger scale (Mann 2005). Farmers typically grow melon, tomatoes, chilis, jicama, and other dietary staples while the field is dormant (Von Hagen and Beltrán 1961).

The entire milpa system, in part because of its low-impact technology, was integrated with ritual traditions. In April, festivities associated with the god of water (Tlaloc) were intended to serve the purpose of bringing rain; the locals made feasts and offerings to the various fertility and harvest gods throughout the year that were directly linked to the milpa cultivation system (Von Hagen and Beltrán 1961). Today, due to religious syncretism, these ritual community celebrations center around the most widely held Mexican symbol, the Virgin of Guadalupe (Sheridan 1988). The legacy of the milpa system is deeper than the persistence of the actual cultivation practices, also persisting in the names of important crops. Nahuatl names for some plants are retained in modern terms, such as: *tomatl* (*tomatillo* in Mexican Spanish, tomato), *ahuacatl* (*ahuacate* in Spanish, avocado), *chocolatl* (*chocolate* in Spanish and English Von Hagen and Beltrán 1961).

The milpa tradition has persisted for more than three thousand years. The *calpulli* system was the form of community land ownership used by the Nahua people under which the milpa system thrived (Von Hagen and Beltrán 1961). The system of milpa is symbolic of social relationships and the universe—it is part of the cosmology of Nahua (and Mayan) peoples found throughout Mesoamerica. The systems of milpa and calpulli were at the center of social organization, which began at the family unit and extended as a web of connections between community members who shared responsibility for an area of land and forest. Governance and decision-making patterns were all rooted in the milpa (Mann 2005).

Archeology of the region is limited, however we do have a few sites of markings and paintings on rocks (including in Telcruz and Cuzalapa). The indigenous language, Nahuatl, is no longer spoken, but lives on in the terms and names of plants, animals, places, and local traditions (some examples of which were given above). This helps to retain the indigenous identity in the region.

Prior to the period of contact with Europeans, who brought horses and facilitated communication with more distant places at quicker speeds, there were existing trade routes connecting the Sierra of Manantlán with other parts of Mexico and Central America. Though little detail is known about exact trade routes or ritual celebrations that would have been cause for travel, we do have oral traditions and some archeological evidence to indicate that there was regular interaction among communities. There was regular trade and communication with coastal communities to the immediate southwest, and connections to the north, in the valley of Ahuachapán (near modern day Autlán).

When the Spanish came to the region, for the purposes of appropriating land for haciendas, maps of the Sierra of Manantlán and existing communities were drawn. One such map still exists and, until very recently, was held by the community of Cuzalapa. (At the time of writing, this map is being cared for and analyzed by university researchers who are using it to try and help arbitrate a local territory dispute.)

Conquest and Colonization (c. 1500-1900)

The Nahua-speaking people who are indigenous to this area suffered major losses in population upon contact. Laitner and Benz (1994) have reported that from 1525 to 1540 there was a 65-78 percent decline in the native population in the Sierra of Manantlán. The Spanish took control of the most fertile lands during this period of colonization. Land and land uses are essential characteristics of the indigenous cultures of Mexico and the Reserve in particular.

The introduction of the *encomienda* system in Mexico (similar to the plantation system of the United States) led to the disappearance of the traditional land tenure systems. Prior to the conquest, rural Nahua Mexico had a formalized land tenure system *calpulli.* The calpulli system of the Nahua people, for example, was based on small kinship groups who managed a marked area of land. Of course, Mexico has always been, and continues to be, constituted by a great variety of indigenous groups. The calpulli system was slowly eroded by the Spanish, and later Mexican, land cessations to the elite class who focused on intensive agriculture on *haciendas* (the functional unit of production in the econmienda system). From 1876 to 1910 the Mexican President Porfirio Diaz supported the growth of haciendas in part to help feed a rapidly growing population (ibid.). The result was that by the end of the Porfiriato (term of multiple offices held by Diaz), about half of the Mexican population constituted a poor peasantry of workers on the haciendas (Wilson and Thompson 1993). Through sociopolitical processes, the many ethnic groups of Mexico have been assimilated into a large peasantry, in spite of the variety of indigenous cultures they represent.

The Mexican Revolution, 1910-1915, was based on the need for agrarian reform and pressure from the rural poor. In 1915, the *ejido* system was created in the Mexican constitution allowing abandoned lands of haciendas to be distributed to the peasants who had been working on them. The Revolution was a success for indigenous agrarian rights in the sense that indigenous people were promised to be restored ownership of hacienda land. From 1915 up through the seventies, waves of agrarian reforms mirrored the social movement and angst of the peasantry (Thompson and Wilson 1994).

Today, there are two primary types of community land tenure in Mexico: *comunidades indígenas* and *ejidos*. *Comunidades indígenas* (indigenous communities) are a traditional land tenure system wherein local people claim an ancestral connection to a specific territory. Establishing the indigenous communities required adequate demonstration of pre-existing corporate entity (a connection to the place) and this criterion gives a degree of added pride to indigenous communities. Indigenous communities have greater flexibility, legally, in terms of how natural resources are shared and divided among community members (Oldfield and Alcorn 1987). They have a unified system for making decisions, but the division of resources can vary according to the internal agreement (Thompson and Wilson 1994; Walsh Sanderson 1984). In some respects, the indigenous communities have greater autonomy.

In its contemporary form, an *ejido* is a combination of the Spanish and Nahua tenure systems. Twenty or more people apply for a land grant and when the grant is approved it is divided into *parcelas* (agricultural plots) among the *ejidatarios* (individuals and their families who constitute the ejido) and grazing land is communal (ibid.). The ejido is the legal entity with elected officials responsible for managing communal grazing lands and external affairs. The President, Secretary, Justice of the Peace, and Council of Members are common governance structures in all ejidos, but some also have a Treasurer. State-based regional offices are the first point of contact for legal matters in ejidos. Later in this chapter, I will explain why the community land tenure systems are important and how they shape and constrain decision-making in the study communities.

Contemporary Period (1900-present)

The 1917 Constitution of Mexico formally abolished the encomienda system, with the promise that an ejido system would be reified and the lands that had been under the encomienda system would be returned to the indigenous communities. The Mexican presidency was dominated by the *Partido Revolu-*

cionario Institucional (Institutional Revolutionary Party, PRI) for seventy-one years, until Vicente Fox, of *Partido Acción Nacional* (National Action Party, PAN) won the 2000 presidential elections. Mexico continued with the conservative PAN party in leadership with the election of Felipe Calderón in 2006, but has recently returned to the PRI party in 2012 (when Enrique Peña Nieto was elected). Since the 1930s, the PRI party has had the greatest influence on economic and social policies in Mexico. Thus, policies reflected the operational structure and goals of PRI members. Throughout PRI's long history occupying the presidency (and other branches of the government), individuals and groups unable to harness strong political power were largely excluded from policy decisions (González-Rossetti and Mogollon 2000).

Post-revolutionary Mexico aimed for improved social and economic conditions. The initiation of the IMSS is attributed to the PRI platform, as it was the brainchild of Lázaro Cárdenas (1935), but the Ávila Camacho administration was the official creator of IMSS in 1943 (Merrill and Miró 1996; Ward 1986; Krauze 1998; Dion 2005). Following the economic crisis of the mid-1980s, Mexico was in need of many economic and political reforms.[2]

The PAN victory for the presidency in 2000 introduced a new dynamic in Mexican politics. The PAN party is more conservative than PRI in supporting a more open free market economy (Dion 2005). PAN's economic policies, combined with its strong relationship with Catholicism and a recent partnership with the Green Party, may have helped PAN gain the support of both rural poor and urbanites (Dion 2005). Among other things, President Fox advocated for substantial increases in health sector spending. Like most promises, any such "substantial increases" were not felt by poor rural families.

The presidency of Carlos Salinas (PRI) began in 1988 and ushered in the use of extreme structural adjustment policies in order to revamp the failing economy and pave the way for Mexico's emergence into the world market via free trade agreements such as NAFTA (González-Rossetti and Mogollon 2000; Krauze 1998). In spite of the stipulated decrease in government spending and the privatization of public firms and services, which were viewed as draining the national economy, the health sector remained largely unchanged during the Salinas administration (Krauze 1998). Since 1988, the health sector in Mexico has been engaged in a process of privatization of healthcare (Harrison 2001). Prior to this, there have been periods of health sector reform (1979-1988), a period of crisis (1967-1979), and prior to that, from 1943-1967, the rapid development of biomedicine and its institutions (IMSS 2005).

Economic troubles surfaced for Mexico again in the mid-1990s under the Zedillo administration. The pension sector of the Social Security Administration was facing imminent bankruptcy, and despite increased taxes the Zedillo administration was working within small margins for reform maneuvers (González-Rossetti and Mogollon 2000; Krauze 1998). Zedillo realized that he could not take on the widespread social sector reforms that Mexico was in desperate need of, so instead focused on stabilizing the nation economically and politically (González-Rossetti and Mogollon 2000). Reforms in the health sector were more reactive than proactive, dealing with lobbyists from the *Instituto Mexicano del Seguro Social* (Mexican Institute of Social Security, IMSS) when they could no longer be ignored (Harrison 2001; PAHO 2003). Though the health sector urgently needed attention, the impetus was not there to make it a top priority during the Zedillo administration.

With the national level economic crises of the 1980s and 1990s as a backdrop, the quality of social services in rural areas went largely unnoticed. As a partial effort to reduce health disparities throughout Mexico, the Zedillo administration established the "National Development Plan, 1995-2000" with two primary objectives (Ward 1986). First, the program aimed to provide access to healthcare services and improve the quality of services provided to rural communities, such as those in the Reserve. Second, the program aimed to promote preventative health strategies, such as immunization and potable water campaigns, in order to cope with the changing epidemiological environment of Mexico.[3] Toward the end of Zedillo's administration, small efforts were being made that had the potential to influence maternal health. For example, in 1997, the Ministry of Health initiated *PROGRESA* (from the Spanish *Programa de Educación, Salud y Alimentación*, or Education, Health and Nutrition Program) to assist impoverished families by giving financial assistance to lactating women and children under five years of age. The Ministry of Health is a major player in implementing these programs and evaluating their impact. This administrative branch has stressed increased accessibility of healthcare services, while improving the quality of services provided. One of the other mechanisms the Ministry of Health uses to improve the quality of services is the regulation and licensing of healthcare providers (UNDP 2003). This component of providing quality healthcare will be discussed in chapter 5 with respect to healthcare sector infrastructure.

In 1999, the per capita expenditure on healthcare in Mexico was US$461 (PAHO 2003). This same year, 53 percent of the total expenditures were private, meaning this money was spent by either employers, insurers, or even individuals (PAHO 2003). The previous year, the out-of-pocket expenditures accounted for only 48 percent of total healthcare spending (PAHO 2003). The PAHO report notes an increasing gap between the rich and the poor, where literacy in rural areas (20 percent) is a fraction of the overall literacy rate in the country (91 percent PAHO 2003). This growing disparity is evident despite "the economic policy [of 1994-2000 which] facilitated a greater expansion of productive activity and employment, channeling greater resources to social spending" (PAHO 2003). Further evidence of the problem are the growing numbers of poor in Mexico (PAHO 2003), which constitute dismal economic outlooks for the healthcare system, and social system in general, of Mexico. Serious action should be taken to promote efficiency and equity to access, especially in rural areas.

The Mexican population can be characterized as increasingly urban. Population growth in Mexico is comparable to many industrialized and developing nations: in 2004 the annual growth rate was 1.4 percent, and the total fertility rate was 2.4 children per reproductive-aged woman (PAHO 2004). Demographically, Mexico is a Second-World developing nation. Public health data also illustrate the status of Mexico as a developing nation. Nearly 10 percent of the population lives below the international poverty line as of 2000 (PAHO 2004). As of 1998, only 87 percent of the population had access to clean water sources, and 73 percent had access to proper sewage disposal (PAHO 2004; World Bank 2005). Over three-quarters of the population (76 percent) is urban-based, and one-fourth of the urban population is concentrated in Mexico City (PAHO 2003).

The trends in Mexico toward urbanization directly and indirectly impact the communities in the Reserve. First, social policies that are aimed at promoting equity in access to social programs and health care are still not evenly distributed to all areas, particularly rural areas. Secondly, the urban areas serve as markets for both the sale of goods and cheap labor by migrants from rural communities. Thirdly, the demographic and health needs of the Reserve reflect, in many ways, those of greater Mexico. In the second section of this chapter, I describe the similarities and differences between my study sample, and its representativeness of the respective communities, in detail.

Ejidos and Comunidades Indígenas

In Mexico today, organized community groups control 50 percent of rural land which are classified as either indigenous communities or ejidos (Wilson and Thompson 1993). The principal difference between indigenous communities and ejidos is in their decrees and access to resources; the history of land tenure and agrarian reform is significant to understanding the processes of decision-making in the Mexican countryside.

Ejidos and indigenous communities across Mexico vary more than they are alike, varying by size, type of natural resources and land, level of technology, and productivity (Thompson and Wilson 1994). Governing institutions in Mexico and the policies they create often fail to take into account this heterogeneity. The variation across ejidos requires flexibility in the policies governing ownership and management of ejido lands, but social scientists have pointed out that this is not normally the case (Wilson and Thompson 1993).

Decision-making regarding the use of natural resources is particularly complicated in areas with special resource management regimes, such as the Reserve. Local political structures are not always effectively integrated with outside governance structures. The Nahua culture can be characterized as based on patriarchal, close-knit kinship communities (Hamilton 2002). This cultural model for decision-making is certainly not unique to the Reserve and continues to be a prevalent conceptual model for local-level governance by community members. That is, most community members defer to elder men when seeking leadership.

Natural resource management and conservation is a process of social action and involves participants at local, national, and global levels. The Zapatista National Liberation Army (*Ejército Zapatista de Liberación Nacional*, or EZLN) has a role in morally, politically, and socially representing the needs of the indigenous and peasant populations in Mexico. They gained international attention in Chiapas in 1994 when they made a dramatic demonstration, seized a town, and demanded the Mexican national government fulfill its promises to the indigenous peoples—particularly in serious agrarian reform that would enable indigenous communities to have better livelihoods (Collier and Quaratiello 2005).

In the Reserve, there are thirty ejidos and five indigenous communities.[4] The poor and marginalized communities have not been a priority at the policy level in post-Colonial Mexico. Natural resources were exploited by local community members but as outsiders came to buy forest resources the traditional means of management became corrupted with the influences of power and money (Rojas et al. 1996). The need to balance conservation and development goals contributes to the complexity (Brechin et al. 2002). In recent years, social movements, such as the Zapatista movement has helped draw political attention to the area as part of the broader struggle for agrarian reforms and indigenous rights in Mexico.[2] Promised agrarian reforms have not been fully realized to date.

One of the goals of agrarian reform was to give the poorest people in Mexico the capacity to produce for themselves. The idea was that if the fertile lands were returned to their native owners, backed by the national bank but managed by the indigenous inhabitants, these communities would emerge from destitute poverty. Technically, the national government "owns" these lands but leaves their management to the local communities. This is an interesting philosophical distinction that becomes blurred in the daily affairs of the Reserve.

In 1960, the Mexican national census reported that 23 percent of Mexico's agricultural land belonged to ejidos. Since the mid-1990s, and as recently as 2004, reports have indicated that more than half (55 percent) of agricultural land in Mexico was organized into ejidos (Staff 1995; CIDOC 2005). The 2004 data indicate that there are 27,664 ejidos and 2,278 comunidades indigenas in Mexico, comprising around 9 million fields and lots (CIDOC 2005).

For most of the twentieth century, ejidos struggled for agrarian reform. In 1991, President Salinas de Gortari criticized the ejido system because the level of productivity in ejidos was not sufficient and certainly not competitive in large markets. In 1992 Article 27 was passed at the federal level, allowing ejidatarios to privatize their parcels of land if they passed a 2/3 majority vote in the Ejidal Assembly. The federal governing body PROCEDE was formed and charged with helping to privatize social property and communal land holdings (i.e., ejidos and comunidades indigenas) and issue small holding certificates, land deeds, and other documentation allowing rental or sale of individually held land (CIDOC 2005). Privatized land is sold or rented—mostly to expatriates or foreign held agri-corporations (Staff 1995). Other program objectives in the privatization of land include increasing the capacity of farmers and land holders to acquire forms of capital for expansion and intensification of production (CIDOC 2005). In the midst of these changes, the declaration of the Reserve created a call for agrarian reform, but not of this sort. The declaration of the Reserve is discussed in the chapter 3 in the context of local governance and natural resource management. The struggle for agrarian reform remains vital and ongoing in Mexico.

Description of the Study Sample

As I showed in chapter 1, my research study included 3 percent of adult men and women in each of the three study communities: Ayotitlán, Zacualpan, and Telcruz. I use the INEGI 2005 data for a rough comparison between the socioeconomic status of study participants and the marginality of the community. I sampled nearly 14 percent of households in Zacualpan, 15 percent in Telcruz, and 23 percent in Ayotitlán, including just over three percent of all adults in each community in my sample (since only one adult per household was included). As indicated in table 2.2, most of the study participants were married at the time of interview; those who reported being single or widowed were most frequently found in Zacualpan (28 percent) whereas being single or widowed was much less common in Telcruz (18 percent) and Ayotitlán (15 percent).

The 2005 INEGI census data is presented alongside the descriptive data of my sample in table 2.3. In many regards, my sample contains the same characteristics as the INEGI data. Households with dirt floors (a proxy for modernization used by other studies in the Reserve; Benz et al. 2000) are represented in the same percentages in Zacualpan and Ayotitlán; however in Telcruz there is a higher rate (68 percent) of households with non-dirt floors than reported in the INEGI census (41 percent). These households in Telcruz also had higher rates of "bathrooms with plumbing"; nearly all (95 percent) of the study sample reporting having them, whereas the INEGI reported only 41 percent. The same trend was found in Ayotitlán, where 82 percent of the study sample reported having a bathroom with plumbing, and only a fifth of the households have it in the 2005 census. This, however, is likely a reflection of inaccuracy in the question. That they have plumbing in the bathroom is true, however observation and direct experience in the field indicates that most indoor bathrooms did not actually function. Yet, in a few ways, the study sample from Telcruz appears to represent households that are slightly better off, in terms of modern domestic technology, than the majority of community members.

In some senses, the study sample from Zacualpan seemed less marginal than the INEGI census data from 2005 suggested. The average number of individuals living in the household in the study (2.63) was much lower than the INEGI rate (4.39). Telcruz and Ayotitlán had similar rates in the study sample (4.61 and 4.49, respectively) and the INEGI census (4.55 and 4.81, respectively), although the Ayotitlán households did appear to have very slightly fewer individuals living in each household on average. Regardless, Zacualpan should not be considered a completely modern Mexican town, as we find there other indicators of the marginal status of the community.

Table 2.2. Demographic description of sampling frame and national census.

	Zacualpan	Telcruz	Ayotitlán
Total Adults (INEGI)	**1724**	**1165**	**683**
Adult Men	815	570	343
18-64	459	325	170
65+	48	32	14
Adult Women	909	595	340
18-64	564	327	161
65+	56	31	15
Study Sample	**54**	**38**	**33**
Men Total	27	21	13
18-35	9	8	5
36-65	9	8	3
66+	9	5	5
Women Total	27	17	20
18-35	9	6	8
36-65	9	6	8
66+	9	5	4
Marital Status			
Married	39 (72%)	29 (76%)	24 (73%)
Living Together	0 (0%)	2 (5%)	4 (12%)
Single or Widowed	15 (28%)	7 (18%)	5 (15%)

Source: INEGI (2005).

Education is a critical indicator of overall socioeconomic status and was measured at the individual level. The census data indicates that illiteracy among individuals over the age of 15 is equally prevalent in Telcruz (13 percent) and Ayotitlán (15 percent). Zacualpan has a much lower rate (5 percent). These results are similar to what was found in my study sample. In Ayotitlán, most (39 percent) had either never attended school or had only attended for the first couple of years (36 percent). In Telcruz, the majority (63 percent) had completed the first couple years of primary education, or more (24 percent). In Zacualpan, only a few (15 percent) had never attended; most of the study participants from Zacualpan had attended primary school (54 percent) or up through secondary

(19 percent). Thus, the only slight difference between my sample and the INEGI census results are that Telcruz demonstrates again a slightly lower degree of marginality in this study sample.

Table 2.3. Descriptive data and census data of study sample for all three communities.

	Zacualpan	Telcruz	Ayotitlán
Number of Households			
Study Sample	54	38	33
INEGI	390	253	142
Houses with Non-Dirt Floors			
Study Sample	48 (89%)	26 (68%)	18 (55%)
INEGI	337 (86%)	103 (41%)	82 (58%)
Average Number of Individuals per Household			
Study Sample	2.63	4.61	4.49
INEGI	4.39	4.55	4.81
House has Bathroom with Plumbing (Septic)			
Study Sample	47 (87%)	36 (95%)	27 (82%)
INEGI	343 (88%)	104 (41%)	29 (20%)
House has Running Water			
Study Sample	52 (96%)	36 (95%)	12 (36%)
INEGI	376 (96%)	200 (79%)	67 (47%)
Highest Level of Education Obtained			
Study Sample			
▪ Never Attended	8 (15%)	5 (13%)	13 (39%)
▪ Kinder/Primary	29 (54%)	24 (63%)	12 (36%)
▪ Secondary	10 (19%)	6 (16%)	6 (18%)
▪ High School	5 (9%)	3 (8%)	2 (6%)
▪ University	2 (4%)	0	0
Illiterate Individuals Over 15 years of age			
INEGI	78 (5%)	153 (13%)	101 (15%)

Source: INEGI (2005).

Any discussion of the role of outside markets and economic activities in Mexico cannot ignore the influences of national and transnational migration. Though my investigation did not deal directly with migration, it is an important domain that overlaps with the economic activity of households and individuals in the Reserve. Whether the individual participant has lived or worked in another place, or someone is sending remittances from another location, the role of migratory work in the Reserve needs to be considered. We also know that cultural knowledge is impacted in its diffusion and dimensions with migration patterns (Berkes 1999).

There are three basic options for migration: rural-to-rural, rural-to-urban, rural-to-international. Davis and colleagues analyzed the roles of social networks in migration decisions (2002). They found that domestic migration for agricultural work was more common among individuals who had lower levels of education, were of indigenous decent, and were from communities far removed from urban areas and culture (Davis et al. 2002). However, migration to urban and tourist areas in Mexico for work tends to be more seasonal and not as highly associated with low levels of education. When an individual is deciding where to migrate, kinship in their social network plays the biggest and most influential role. Yet, interestingly, ejido-level networks in the United States and Mexico act as a type of substitute for migration to the United States and instead complement domestic migration (Davis et al. 2002).

In the Reserve, it is not uncommon for a family to have an immediate family member or close friend working in the US. Yet, for this investigation, data directly measuring migration and its impact on the household economic activity profile were not measured. The census data indicated that Zacualpan was the only of the three communities that had individuals above the age of five living in the United States—they reported a total of thirty seven individuals (INEGI 2005). In the future, it would be very beneficial to examine the impacts of migration and migration networks in the emerging social and economic activity profiles in the Reserve, particularly in relation to perceptions of nature and space.

There are important ways that individuals interact within and beyond their communities. In this book, I have relied heavily on the household as a primary unit of economic, social, and cultural analysis. Within a household there are shared values and resources, but once we look outside the household there is variation in the types and meanings of interactions with other community members.

Social welfare programs are a very important component of the household economic stability in the Reserve, and also a very prominent social issue. A former employee at the Management of the Reserve explained that, "in Ayotitlán—similar to Cuzalapa—they are all lazy and dependent on government handouts. They don't want to work." Of course, his perspective was biased because he has at other times noted that he is not a very "social person." But, regardless of the prejudice that this statement embodies (and it was a very common explanation for the current status of people in the Reserve), there is a sentiment that, "in Ayotitlán—in contrast to other parts of the Reserve—they are dependent on the government handouts." When I asked one health services provider what, then, could be done to help improve the socioeconomic conditions in Ayotitlán, he responded that, "they need better roads and access to television. Access to the outside world would help them want more things, and thereby motivate them to work more." Again, the idea is asserted that people are poor because they are unwilling to work.

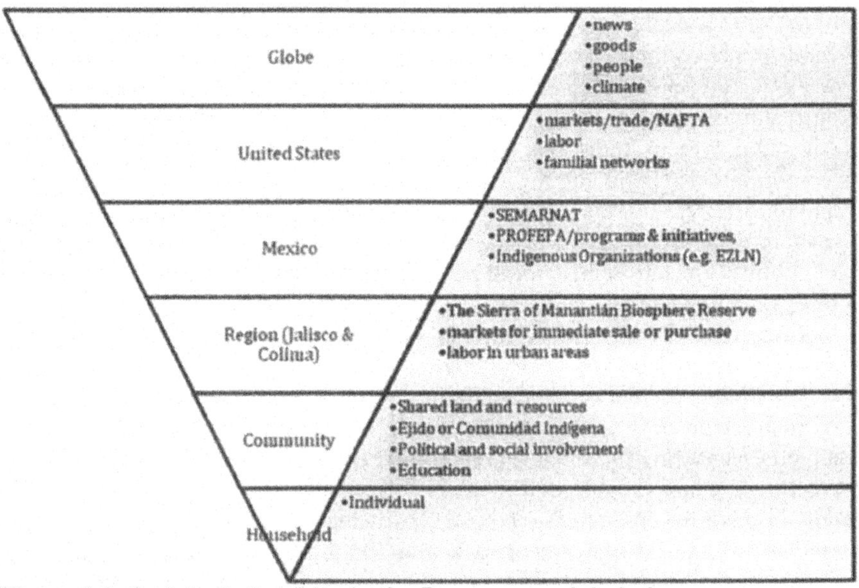

Figure 2.3. Social relationships in the Reserve—a global view.

People in Ayotitlán, at the same time, feel frustrated and trapped. They are prohibited from exploiting their natural resources, and they perceive that others are benefiting from their land and forests. There is bitterness when they talk about how "outsiders come and take their study and information and then they leave." Because their lands and forests have become the laboratories of so many people they expect to receive the royalties of that work. So, what others perceive as a dependence on government handouts, the people of Ayotitlán say is very little pay for the work they are doing, they are "The owners and stewards of *milpilla* [the genetic ancestor of maize]."

Alloparenting is common in the Reserve, in part because many of the biological parents migrate for labor. The household is commonly constituted by the parents, their daughters, and grandchildren. The spouses or boyfriends of older daughters (in their twenties or thirties) very often live with their own parents. Two main reasons that young adults do not live apart from their parents: (1)

costs of maintaining a household apart from the parents, and (2) constant migration for work and instability of work. Hence, alloparenting and pooling economic resources—and risks—is a very wise strategy.

One of my key informants was raised by her grandmother in the Sierra of Manantlán. The mother of my informant had gone to Los Angeles to work where she got pregnant, but she came back to the Sierra of Manantlán to have her baby (mostly because she "did not want to try to raise a child by herself in the United States") and returned to Los Angeles, leaving my informant with her grandmother who raised her.

There was a Convocatoria de Comunidades Indigenas (Conference of Indigenous Communities) held in Mexico City in August 2008. It was convened by the Comisión Nacional para el Desarrollo de los Pueblos Indígenas (National Commission of the Development of Indigenous Peoples; CDI). The community of Ayotitlán sent about three adults and two or three of their university students. The costs of attendance (Mexico City is about a fifteen-hour drive from the Sierra of Manantlán) would be put together through the Consejo de Mayores. They were soliciting contributions from community members to help pay for the gasoline, meals and lodging, and for a vehicle that they could travel in.

At the IMECBIO, the researchers are keenly interested in engaging with community members in addressing sustainable development. However, one major hurdle to working productively in the Reserve is, "the fact that communication channels between the Management of the Reserve and IMECBIO are seriously lacking." In both the Management and the IMECBIO there is acknowledgment of the tensions between their institutions. A former Director of the Reserve commented that the researchers at IMECBIO "think we are here to care for their laboratory" referring to the Reserve. These territory battles inhibit productive relationships of cooperation and collaboration; community members are not only aware of these tensions but take this into account when deciding on aligning themselves with either institution.

Local Cultural Identity

Material Culture

Students who recently graduated from a Mexican university and have to perform their social service staff local schools in the Reserve. There are also, occasionally, full-time teachers. In Telcruz, a young woman was working at the elementary school. She commuted during the week from Autlán. The middle school (or junior high) and high school students had to commute to nearby communities to continue their education, which few did. In Ayotitlán, there were three full-time teachers in the elementary school and at least one full-time teacher in the middle school. For high school, students had to commute to Cuautitlán (which took at least one hour each way by public bus). In Zacualpan, the kindergarten and elementary school had two or three full-time teachers who commuted in from Colima. After elementary school, students continue their education in nearby Valle de Alvarez (a suburb of Colima).

In Ayotitlán there is a *Casa de Culturas* (Cultural Center) where a small library, museum, and computer lab with internet access are located. There are two desktop computers with internet access, a printer, and a small room that houses some brochures of previous projects in Ayotitlán (including copies of a handful of manuscripts regarding the history of the region and milpilla). There are also a few archeological artifacts on shelves—some broken pottery and a few stone tools. The archeology of this region is seriously deficient, and the few items that have been found were reported by farmers in their fields or near the mine. The facility sits adjacent to the large community hall where assemblies are held; it is on the main road that provides access to Ayotitlán. Funds for constructing the facility were provided by the CUCSUR, which also provides most of the money to maintain the site. During the week, a young woman staffs the location. She chatted online via Windows Messenger while talking to me about the failed traditional medicine project in Ayotitlán. She told me that circa 1999 a traditional medicine house, or clinic (*La Casa de Medicina Tradicional*), was constructed in town behind the church. Some medicinal plants were grown in the clinic gardens and a few local healers staffed it. But, due to community conflicts regarding the use of the facility, it has been unused for years. In 2007, it was being used by the local church as a storage closet. Over the coming months, I continued to investigate the story that the young woman shared with me and I elaborate the details of the traditional medicine house in Ayotitlán in chapter 9.

An exchange student at CUCSUR worked with community members in Cuzalapa to document medicinal uses of the tomato (principally) and other food-plants in Cuzalapa. A local key informant worked closely with the student (Sarah Brown) and together they interviewed people in the community—many of whom have since passed away—and asked them how they prepared different home remedies. They prepared a report of all these uses as part of the way to retain that cultural knowledge. This product was not published widely, but there are copies in a few key places of Cuzalapa, including the Cultural Center and the offices of the local government. This is an example of a research project where the end product was tailored to community members—the primary audience was not academic nor outside the community.

Tradition and Celebration

The Council of Elders helps arrange community celebrations in Ayotitlán. They solicit volunteers for hosting the party and others to help provide food and beverage refreshments. There are celebrations in Ayotitlán commemorating Mexican Independence in mid-September (the actual Independence Day for Mexico falls on the 16, but the celebration will be on the weekend nearest this date). Easter, particularly the entire week leading up to it which is called *Semana Santa* (Holy Week) is a huge annual celebration in this part of Mexico. Celebrations begin on Friday afternoon and last through Sunday. There are numerous Masses each day (usually three or more each day), family gatherings, and much music and drinking. Virtually all of Mexico has the week off from school or work, or at least a shortened work week, and it is common for families to take a vacation. During this week, ecotourism sites in the Reserve received more visitors, particularly the new Riverside Park in Zacualpan.

Community festivals and celebrations revolve around predominately Catholic religious themes. There are only hints of Nahua cultural traditions. At one event, there was a group of traveling Catholic Huichol. There were three adult leaders who played instruments and a group of young girls and boys (about ten of them) from an orphanage. They sang Catholic songs and played a small guitar and use bongos. Their participation struck my friends as odd, so they asked me (the cultural anthropologist) to explain whom these people were and why they were at a community celebration in Ayotitlán. After some investigating, I found out that they were invited by the Council of Elders to participate; they were on a traveling mission "headed toward Chiapas spreading the message of Jesus Christ's love."

It is always best—efficient—to celebrate numerous things on one occasion. Thus, at any particular community celebration there are likely to also be birthday celebrations, anniversaries, and baptisms. On one special occasion, a friend's graduation from the University of Guadalajara, South Coast (CUCSUR) coincided with the celebration of the Virgin of Guadalupe in Ayotitlán. The dates are not fixed for these celebrations, which always fall on a Friday, Saturday, or Sunday. In this particular region of Mexico, hitchhiking is still an acceptable—indeed, common—mode of transportation. Bus routes are limited and too costly for many of the people who live in the Reserve. We made our way to Ayotitlán Friday morning, arriving in the heat of the afternoon in time for the lunch feast being prepared. Though the hot sun was baking down on us as we climbed the sandy dirt road to my friend's house, we hiked on. Two or three chickens had been killed to prepare the broth soup, accompanied by plenty of thick, handmade corn tortillas. Hunkering down in the shade, we picked at the chicken bones served with the broth and gulped large glasses of sugary hibiscus iced tea. As we were eating and drinking copious amounts of food, the mother and uncle of my friend prepared the cow that had been killed for the evening's festivities. We went down to the river to bathe and rest prior to the big celebration began. There we splashed in the knee-deep water, washed, and laid ourselves out on the large rocks to dry.

On our way back from the river the festival was already starting. Hiking back to the village, we heard the banging on the large drums—monotonous, bass drums, broken by the rapid spattering on tabor drums. They brought the four *imagenes* (figurines, or symbolic images) of the Virgin of Guadalupe and local patron saints down to the abandoned health outpost in front of my friend's house. There were decorations and everyone joined in by lighting candles and paying respect to the images. The images are ornate dolls. Fragile, they are carried in glass housings that are open only at the front where the pious dolls are adorned with flowers and candles. My friend remained in the rear of the crowd observing and making sure her sisters served refreshments to everyone. Realizing that this was the highlight of the party for my friend (parading the figurines) I offered to take a picture of her with her sisters at the party. She bashfully refused. I could see the pride and sense of fulfillment on her face as she graciously hosted the community celebration.

The Virgin of Guadalupe has a very profound influence in Mexico. Being a native woman who appeared to an indigenous man in Mexico—her image remains a symbol of the mass conversion of indigenous people in Mexico. Indeed, her apparition catalyzed the conversion of millions of Mexican natives to Catholicism in the 1500s (Wolf 1958). She is a syncretized symbol, embodying the Virgin Mary and the indigenous Nahua goddess *Tonantzin* (Goddess of Earth, representing fertility). The Virgin of Guadalupe is undoubtedly the most significant cultural and religious symbol of Mexico and serves as a vivid reminder of the *mestizaje* (mixture of European and indigenous) of cultures.

The variety of music is an interesting aspect of a celebration in the Reserve. There are generally two types of musical groups present: traditional drumming troupes, and *bandas* (bands) that play the regional music called by the same name, banda. Banda is a loud, obnoxious, bouncing music associated with drinking and rowdy dancing. It is found predominately in the southern Pacific coastal regions of Jalisco and is distinguished from Durangense (from Durango), and Norteño (from the border states), which contain similar musical and social attributes. The traditional drumming troupes come in two teams, red and blue. The red group symbolizes Spain and the blue group Mexico. They act out symbolic battles at various times during the two or three days that they play (night and day).

There are two types of music, then, that are present at community celebrations in the Reserve. Traditional music groups are percussion teams (almost always all men, although I did see one young lady playing with them once) of about five to ten players and a leader. The leader of the percussion troupe calls out the songs and also leads the chants. They move throughout the community, up and down the central streets, making stops at the homes of predesignated community members. The percussion troupes lead the procession of community members and, most importantly, the images of the Virgin of Guadalupe. They stop at the homes of families who have volunteered to help host the party by offering refreshment for the party—of which there are usually between fifty and seventy-five men, women, and children. Some homes, where they are able to afford the extra expense, prepare large batches of steamed tacos (filled with refried beans or cooked potatoes), rice, homemade cheeses, or other seasonal *botanas* (snacks). All homes offer a flavored, sweet beverage to their guests. In the evening, the beverage of choice is coffee with cinnamon, or simply hot cinnamon tea. Guests are likely to augment their drinks with cheap, strong sugarcane alcohol. (This is noted in spite of "Dry Laws" that have been instituted in Ayotitlán to help curb drunken-disorderliness on celebratory weekends.)

The main host of the party, in this event my friend who had just graduated from college, is responsible for arranging the musical entertainment and paying for it. The traditional musical troupes, one blue and one red, are the most important for the ceremonial ritual. They lead the procession of the images through the town and to the designated homes of sponsors. The Father performed a special Mass toward the end of the evening, when the images of the Virgin of Guadalupe were brought into the *Templo* (Catholic church).

After Mass, the party continued with other snacks (*duros* are a favorite fried chip), coffee (with liquor, in spite of a dry law), and a variety of offering to the hosts (my friend and her immediate family). He thanked them specifically for the donation of the large cow, the music, and other gifts they had made to sponsor the party. Late in the evening the blind man who is the caretaker of the images spoke loquaciously about the gratitude for the support and the needs of the community that would be met because of their wonderful care for the images. There is a mayor in each locale and I wonder if his lack of sight/vision is somehow associated with him gaining this responsibility. Liquor, *tortillas de queso* (homemade cheese tortillas), and *picadillo* (shredded meat with spices) were among the things gifted to our friend the host, who shared her spoils with us.

Celebrations in Zacualpan reflect more urban, contemporary Mexican traditions. For example, there are no traditional drumming troupes but only banda and mariachi. They have a more urban community, with a bullfighting stadium where they hold rodeos (*jaripeo*). Rodeos are the main social attraction, and are noticeably divorced from the procession of the images of the Virgin of Guadalupe.

Notes

1. *Nahua* refers to one of the largest groups of indigenous peoples in the Americas, and are commonly known by the name "Aztec". The Nahua language group includes the group referred to as "Aztec" in common parlance. There are regions of modern-day Mexico where the Nahua language is still spoken, however many of the indigenous descendants of the Nahua no longer speak the language. The cultural and religious traditions of the Nahua are still practiced in varying degrees across Mexico.

2. For further discussion of the political and social processes related to the development of IMSS during the 1960s, 70s, and 80s, see Ward (1986); and for the period from the 1917 Constitution to the 1990s, see Harrison (2001).

3. Issues of tenure, privatization, agrarian reform in Mexico, and globalization are all implicated in these processes (Thompson and Wilson 1994).

4. In this book, I use the term "community" in the general sense, but when making a specific distinction between the two types of corporate communities revert to the terms "indigenous community" and "ejido."

Chapter 3
Conservation and Subsistence Strategies

This chapter describes the ways in which the people of the Reserve make a living. Their use of natural resources is implicit in these processes, but is regulated and shaped by the presence of management policies and a special tenure regime. Relationships with the natural environment are critical to understanding how people retain their culture and manage shared forest resources.

First, I describe the agricultural and other uses of forest resources and products in the Reserve in the context of the diversity of the natural environment. The exploitation of forest agriculture is tantamount to understanding the culture of Nahua people in the Reserve. Next, I continue with a detailed quantitative description of the use of forest resources through these primary activities of agricultural and animal husbandry. This describes the ways that conservation and development programs may impact the ways in which community members seek to sustain their households and increase their economic opportunities.

Finally, I consider some of the key conservation-oriented goals in the Reserve. I explain the key concepts of sustainable development and conservation, which remain focal activities in the Reserve at the time of writing. I begin with background information regarding the goals of sustainable development, including the theoretical foundations for the types of initiatives found in the Reserve. Later, in chapter 6, I provide specific examples of recent, and ongoing, projects in the Reserve. The context of development is essential to appreciating economic activities of individuals and households in my study.

Natural Resources Usage in Study Communities

The Reserve continues to struggle with a variety of environmental and social concerns, to which researchers have been calling attention. Forest fires that are not properly prevented nor controlled persist. Changes in the characterization of land uses have continued the degradation of the local environment. The extensive livestock grazing is not well regulated. Recent years have seen the introduction of underground black markets, and the production of illegal drugs (namely marijuana). These problems, along with contamination of the rivers and streams by industries upstream (mainly the sugarcane processing plant), have prompted a stronger series of governmental interventions. The mere designation of the Reserve as a protected area in 1987 did not mean that local guidelines and enforcement were in place; it was not until 1994 that true protection of the Reserve began to make a change in the way that the area is managed.

After the discovery of teocintle, and the international attention that the Sierra of Manantlán received, President Miguel de la Madrid declared it a biosphere reserve in 1987. The goal was to promote environmental conservation and social development (following the Mexican model of a biosphere reserve). The Mexican model of a biosphere reserve is based on the principles of sustainable development and conservation strategies centered on community. The idea of a biosphere reserve is somewhat different in Mexico, having been labeled the "Mexican Model" wherein the main objective is to incorporate local institutions and people in conservation of the local environment (Graf et al. 2003; Jardel 1992; Halffter 1995).

Since the discovery of the genetic ancestor of corn, the Reserve was created, accompanied by a proliferation of biological and environmental research in the area. The discovery of the genetic ancestor of maize in the Reserve led to its original formation and there has since been a proliferation of biological and environmental research in the area. The national declaration of the Reserve cites the central purposes to be (Graf et al. 2003; Jardel 1992):

1. Territory demarcation and protection
2. Sustainable extraction and management practices
3. Scientific research
4. Generation or reintroduction of appropriate technologies
5. Education about the environment
6. Formation and capitalization on human resources
7. Fortification of local organizations

The focal sustainable development activities in the Reserve are centered on: the controlled extraction of forest resources; management of agricultural lands and livestock; and ecotourism. In this book, I address the later two and not the extraction of forest resources, such as timber, because it is not a prevalent activity in the three sample communities. Land use for agriculture and animal husbandry is essential to understanding the livelihoods of individual households in the study sample, and ecotourism is an emerging source of economic activity for many individuals in the Reserve.

In the three communities Zacualpan, Telcruz, and Ayotitlán, there are characteristic differences in how forest resources are used. In general, there are four categories of natural resource exploitation in the Reserve: (1) non-intensive agriculture, (2) animal husbandry, (3) use of timber, and (4) use of non-timber forest resources (Graf et al. 2003). The management of land and resources is also determined by the delimitations of core and influence zones of the biosphere reserve, and specific regulations of the Management of the Reserve. There are three zones, as indicated in figure 3.1, and none of the study communities are located in the core zone. Though the populated areas of the communities studied do not lie within the core zone, many of the field and other forest resources used by families in Ayotitlán do come from the core zone. Furthermore, the populated center of Zacualpan is technically just outside of the Reserve boundary, however the fields and other natural resources used by people in the community lie within the Reserve, so it is common practice to treat Zacualpan as part of the Reserve.

The core zone is reserved for conservation and the activities within it are highly restricted. The zone of influence is the periphery of the Reserve and denotes the environments immediately associated with the activities in the Reserve. Zacualpan, for example, lies just outside of the official boundary of the Reserve, however most of the fields belonging to people in Zacualpan are within the Reserve limits and so this community is highly impacted by the Reserve management policies. Management of the Reserve and enforcement of conservation policies are the responsibility of field representatives of the Management of the Reserve. These officials ensure that the local needs for economic and social development are being met in a way that is sustainable and help to conserve environments. Principal economic activities involve the environment and are also central to local culture: agriculture and animal husbandry.

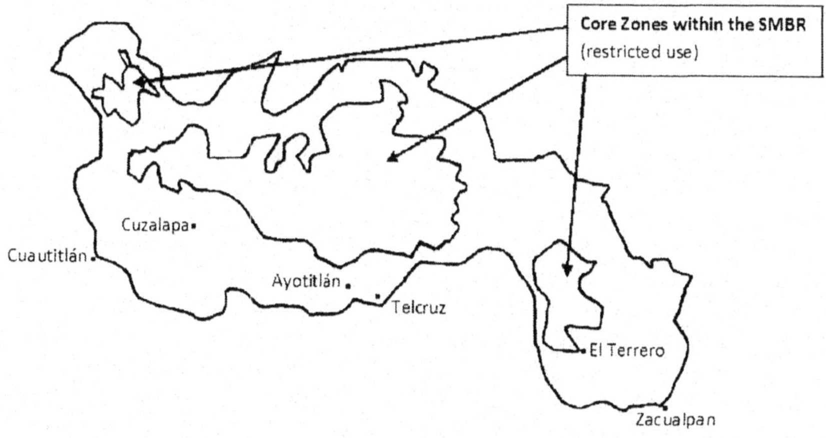

Figure 3.1. Map of the main study communities and the designated core zone of the Sierra of Manantlán Biosphere Reserve. Copyright E. A. Olson.

Economic Activities

Most families still have a mixed-subsistence strategy: some agriculture, some animal husbandry, and some waged labor. Together, these different components help provide income security for the household. The influence of waged labor is low in all communities except Zacualpan. In all communities, government stipends are a major source of cash income security—but many households neglected to report this as part of their income.

There are three types of paid labor prevalent across my study communities: (1) migratory waged labor in the United States or other distant location; (2) paid labor in an urban area near the Reserve; and (3) paid labor in the Reserve. The latter is very uncommon and represents only those who have local businesses, such as a store, or work as paid representatives of the Management of the Reserve for specific projects. No household relies strictly on one form of income or economic opportunities. Most have varying combinations of agricultural or animal husbandry activities, paid labor, and other sources of support (such as governmental stipends). None of the households included in the study reported economic activities that do not fit under these categories.

Agriculture

The count and type of agricultural fields is shown in table 3.1. In Zacualpan, nearly one third of households surveyed (28 percent) do not have a single field, and of those fields that there are in Zacualpan the majority are described as agricultural (59 percent). Telcruz is not necessarily the most agricultural of the three communities, but all of the households surveyed there did have at least one field, and the fields are mostly agricultural (47 percent) or secondary vegetation and low mountain (32 percent). Both of these communities stand in contrast to Ayotitlán where over half (52 percent) of the households surveyed reported not owning their own field, but the majority in Ayotitlán do engage in agricultural work as a primary source of income.

Definitions of land tenure change across communities, and in Ayotitlán when asked how many fields the family owns, there are two reasons why a family might respond that they do not have any. First, because they do not view the field(s) they work as their sole responsibility, but rather as a shared resource among multiple extended family members. In this instance, they would not want to report the field as being their sole property because doing so would not accurately account for the shared nature of the resource. The other primary reason a family might not report having a field is that they are not working fields that are theirs, but rather working as hired hands on fields belonging to others. Nearly all (91 percent) of the households surveyed indicated working in between one and three fields over the last year. This is further complicated by their designation of the majority of their fields as "agricultural" (79 percent). This is actually a misnomer; the fields they use for agricultural purposes are technically classified as "agricultural" by the management of the Reserve, thereby allowing them to work these fields annually. They are not allowed to simply clear fields whenever nor wherever they choose. There are regulations concerning how many years a field must go to fallow, when a new field can be cleared, and so on. Thus, their response to this question reflected their knowledge of management regulations on land use. Based on observation of the area, most fields are actually secondary vegetation.

Table 3.1. Agricultural fields and animal husbandry by study community.

	Zacualpan (n=54)	%*	Telcruz (n=38)	%*	Ayotitlán (n=33)	%*
Count of fields held						
0	15	*27.8*	0	*0.0*	17	*51.5*
1-3	39	*72.2*	37	*97.4*	16	*48.5*
4 or more	0	*0.0*	1	*2.6*	0	*0.0*
How many fields/plots did they use in the last year?						
0	20	*37.0*	1	*2.6*	3	*9.1*
1-3	34	*63.0*	36	*94.7*	30	*90.9*
4+	0	*0.0*	1	*2.6*	0	*0.0*
Mean Field Size (Hectares)	1.24		1.58		1.88	
Median Field Size (Hectares)	1.00		2.00		2.00	
Primary field type						
Agricultural	32	*59.3*	18	*47.4*	26	*78.8*
Secondary Vegetation, Low Mountain	7	*13.0*	12	*31.6*	2	*6.1*
Forest, or Deep Mountain	0	*0.0*	7	*18.4*	3	*9.1*
Not reported or missing	15	*27.8*	1	*2.6*	2	*6.1*
What did you grow in this field?						
Corn	23	*42.6*	34	*89.5*	25	*75.8*
Milpa or mixed (corn and others, usually beans)	12	*22.2*	1	*2.6*	6	*18.2*
Plums	1	*1.9*	1	*2.6*	0	*0.0*

Nothing, or they did not plant	18	*33.3*	2	*5.3*	2	*6.1*
Where did you get the seeds?						
N/A	18	*33.3*	3	*7.9*	2	*6.1*
Already had	19	*35.2*	21	*55*	26	*78.8*
Already had some, and bought some	2	*3.7*	0	*0.0*	0	*0.0*
Were purchased	9	*16.7*	13	*34.2*	5	*15.2*
Were gifted	6	*11.1*	1	*2.6*	0	*0.0*
Count of milking cows						
0	37	*68.5*	34	*89.5*	30	*90.9*
1-5	15	*27.8*	3	*7.9*	3	*9.1*
6 or more	2	*3.7*	1	*2.6*	0	*0.0*
Mean Total Cows, Bulls (all livestock)	4.46		1.34		1.24	

* Percentages have been rounded-up to the nearest tenth, resulting in some small anomalies in some of the categories.

The average field size is greatest in Ayotitlán (1.88 hectares) and smallest in Zacualpan (1.24 hectares), while Telcruz lies between the two (1.58 hectares). This difference between the communities is not significant ($p=0.152$). Milpa fields are mostly found in Zacualpan (22 percent) and Ayotitlán (18 percent) but in all communities, corn is the most common crop: Zacualpan (43 percent), Telcruz (89 percent) and Ayotitlán (76 percent). These data are all summarized by community in table 3.1.

Lastly, it is important to discuss the source of seeds used by farmers in the Reserve. Telcruz households either already had (55 percent) their seeds, or purchased seeds (34 percent). In neither of the other two communities are seeds purchased as frequently: 17 percent in Zacualpan and 15 percent in Ayotitlán. (The only major exception to the importance of where seeds are obtained is evidenced by the 33 percent in Zacualpan who did not need seeds—either because they did not perform any agriculture or because they were harvesting perennial plums off of trees.) Of course, it is most common to use seeds retained from the previous crops (35 percent in Zacualpan did this, and 79 percent in Ayotitlán). Conserving biodiversity and minimizing harmful environmental impacts is largely related to the source of seeds. Purchasing transgenic seeds, for example, greatly increases the total crop size but initiates a dependence on purchasing the seeds and chemicals used to grow it. The costs associated with producing the increased yield are substantial; when corn market prices are high this pays off nicely for the farmer. When corn market prices waiver (as they have over the past few years in Mexico due to things like imported corn surplus from the United States), these farmers are in a crisis. Thus, using the milpa system is not sustainable at a large-scale level. Neither is using transgenic and intensive agricultural techniques for the production of Mexico's staple crop: corn. Here, in the Reserve—the birthplace of corn—the means and modes of production draw the essential lines of local culture and daily life.

Animal Husbandry

Cattle are both culturally and economically important in Mexico. The traditions surrounding bullfighting (originally introduced by the Spanish colonizers) continue to be a central aspect of Mexican cultural traditions, particularly in Jalisco. However, the economic benefits of cattle ranching are equally important and are a symbol of wealth and affluence. Table 3.1 summarizes the types of cattle and milk cows held in the three study communities. Very few households in Ayotitlán have any cows (9 percent) and the average total number of cattle and cows held is 1.24. In Zacualpan, on the other hand, there are an average of 4.46 total cows and cattle held, but of course these are held by a third (32 percent) of the study sample. Only 11 percent of households surveyed in Telcruz have any cows and the average total there is comparable to that of Ayotitlán (1.34).

Conservation Programming

Oldfield and Alcorn have strongly argued that traditional knowledge and agroecosystems should play a central role in conservation and development programming (1987). They refer to "traditional agroecosystems" as the set of farming techniques and practices used, including the use of land and forest resources, which form the basis of the local farming lifestyle that is situated in a specific time and place (Oldfield and Alcorn 1987). Many researchers have asserted that conservation requires the integration of traditional ecological knowledge (Berkes et al. 1995; Oldfield and Alcorn 1987).

There are many benefits to conserving traditional agroecosystems, particularly since the living world contexts that conservation is meant to impact cannot be replicated in a laboratory or greenhouse. Ecosystems are dynamic and it is necessary to consider the co-evolution that has gone on and continues to do so. Freezing seeds and samples in a laboratory is not sufficient to truly understand ecosystem dynamics. Weeds and wild species co-evolve with domesticated crops and are integral to the system.

Cultural and socioeconomic factors are intertwined with the agroecosystem, specifically in terms of the ways that humans have coped with environmental conditions and modified landscapes. Seeds are essential in the process of cultivation and technology. The means by which they are selected, shared, and stored will all be important factors that get lost if the traditional agroecosystem is not conserved alongside them. Livestock and other harvesting of wildlife species are integrated into the schedules and influence the shapes of habitats. These are all culturally shaped aspects of living in forests. Much intracultural heterogeneity in terms of experimentation and individual variation operates within a shared cultural context (Gerritsen 2004). Trade systems are essential to broader social relations and economic relations that impact ecosystems. Peasant farmers are very vulnerable to the risk of using new seeds and intensive agricultural technologies because of the high investment requirements and the fact that these marginalized communities are usually found in unstable social and political environments. The traditional agroecosystems are rarely engaged in monocropping, instead prioritizing integrated forest resource use and other minor crops that are for food, medicines, or other application.

The UNESCO Man and the Biosphere (MAB) provides a designation of biosphere reserve areas that espouses the rhetoric of integrated wilderness conservation and rural development programs. MAB recognizes a biosphere reserve, but it is incumbent on the national government to appoint an agency that directs and manages the activities of the reserve, modifying MAB guidelines as necessary to meet the local needs (Halffter 1994; Oldfield and Alcorn 1987). These programs fit under the designation of Integrated Conservation and Development Projects (ICDPs) which are explored more in chapter 6. The ICDPs require long-term support because developing nations do not have the money nor local research power to adequately accomplish the goals of local traditional agroecosystem conservation (Oldfield and Alcorn 1987). These ICDPs are aimed at integrating research, biological off-site research facilities, and community-based agroecosystems that capitalize on traditional technologies and management practices (Oldfield and Alcorn 1987). A new research program is being undertaken by investigators at the IMECBIO that fall under this rubric regarding the milpa cultivation system in the small community of San Miguel (within the ejido of Ayotitlán).[1]

It may seem obvious, but for many years it was overlooked that biologists simply cannot handle an ICDP on their own—these need to be interdisciplinary and multifaceted programs. Social scientists are needed to research the local technologies and practices. Another major critique of ICDPs such as these is that they are too costly to be realistically feasible (Oldfield and Alcorn 1987). Peasant farmers, living in developing areas like the Reserve, will certainly want the opportunity to incorporate intensive agricultural techniques and improve their socioeconomic status. It is necessary to provide them with a supplement as an incentive to avoid mono-cropping and intensive, fertilizer and chemical-based agriculture. Providing supplements is a dangerous road: this road leads to unrealistic expectations, very rapid changes in socioeconomic status, and has a powerful effect on local sociopolitical structures (sometimes even doing more harm than good (Oldfield and Alcorn 1987). Nevertheless, there are some indigenous groups whose members are looking to conserve their cultural heritage, and these communities are ideal candidates for ICDPs.

One of the most commonly cited and optimistic mechanisms promoting conservation of traditional agroecosystems is the following three-pronged approach:

1) *In situ* integrated conservation and economic development program,
2) *In situ* conservation research sites, and
3) *Ex situ* reserves. (Oldfield and Alcorn 1987)

In marginalized areas, local varieties are much more dependable, reliable, and accessible. Peasant farmers simply cannot access transgenic seeds and the associated technologies needed to be successful at farming them (Oldfield and Alcorn 1987). In any case, with all ICDP it is absolutely essential to work with local farmers in shaping the goals of how their agroecosystems are changing and evolving, so that the farmers themselves continue to be owners of their agroecosystem. Reserves cannot act as "cages" but depend on actual buy-in from community members to accomplish an integration of traditional and scientific knowledge systems in the promotion of traditional agroecosystem conservation (Oldfield and Alcorn 1987). However, this is only one type of ICDP with conservation goals—there are other approaches that are likewise relevant in the Reserve.

The expressed goals of two key outside stakeholder institutions (IMECBIO and the Management of the Reserve) are complimentary and congruent with the ideals of ICDP. The IMECBIO is an institute created in 1985 by and operating within the University of Guadalajara. At IMECBIO the pedagogy centers on scientific research and its application within the research communities (Martinez et al. 2006). The bulk of the scientific research in the Reserve pertains to biodiversity of plants and crops. This information is collected in the fields and forests belonging to the community members. While many of the actual field researchers are students from the communities who study at the regional campus of the University of Guadalajara, the lead investigators tend not to be indigenous peoples from the Reserve. Students from the Reserve who study at the local university (to which the IMECBIO pertains) are enrolled in various programs concerning natural resource sciences. These students, being from the Reserve, are able to conduct research within the Reserve without going through the community consent process. Meaning, they are allowed to carry out surveys and collect plant specimens without obtaining any special permits. Having these students on research teams may help facilitate positive community relationships because the community members can more easily identify with the research being done. Additionally, these students may bring a specialized expertise and knowledge of the area that contributes to the research being conducted.

Due to the technical expertise and resources of the IMECBIO, they are an important component of the sustainable development process in the Reserve; they collaborate and act as accessories to many of the forest management issues for the Management of the Reserve (ibid.). Both institutions share the same conservational and sustainable development goals—and they are able to complement the specific objectives of each other. The communities within the Reserve are hubs where these two stakeholder institutions interact. In general, the IMECBIO provides researchers and scientific research, which the Management of the Reserve then applied to its administration of the Reserve.

A variety of research and developmental activities are undertaken in the Reserve. Success of these activities is measured in different ways: at IMECBIO the production of scientific data is the key output, and at the Management of the Reserve the realization of community projects is the primary measure. Scientific ecological studies are predominately used to monitor and track biodiversity, having direct application to conservation in the Reserve. The implementation, and continuation, of sustainable development projects is predominately the work of the Management of the Reserve, who focus on two key outcomes: monetary investment from grants or other agencies and qualitative assessments of community engagement. Community capacity building is seen in the continuation and growth of projects; improved socioeconomic conditions and access to resources have tangible benefits. The impact of scientific research is less direct and tangible. Furthermore, the different types of impacts should be expected on different timeframes and the long-term results are most essential for true sustainable development and conservation. Thus, there are different ways of measuring "success" within the Reserve and we may not yet have actually seen the complete impact of some of the activities that are underway.

Currently, the Reserve is characterized by an increasing dropoff in agricultural surface area. The import of cheap mono-cropped products from other countries (especially the United States and Chile) has decreased the profit margin for local producers. Low market prices for goods produced and high costs of intensive agricultural techniques leads farmers to switch to animal husbandry (when possible) and seek cheap paid labor in nearby urban areas (Graf et al. 2003). In other areas of Mexico and the globe, farmers will sell their parcels of land to large, often foreign owned, agribusinesses. Increased usage of intensive agricultural techniques, such as mono-cropping and heavy dosing of chemicals (both fertilizers and pesticides) result from the process of agribusiness. The cyclical effects are felt in an increased decrease in the quality of soils, in turn demanding more intensive agricultural techniques. And so the cycle continues.

The reduction in agricultural productivity is also associated with the use of land for livestock tenure, which has increased alongside the decline in agricultural productivity. Importantly, cattle have a higher cash value in the market (and the prices for cattle have been more stable in recent years than those of agricultural products such as maize or tomatoes). Many of the communities continue to designate communal grazing areas (Graf et al. 2003). These, along with government credit programs to purchase cattle, have contributed to the increase in livestock tending.

Prior to the formation of the Reserve as a protected area, lumber companies extracted without permission and preyed on local farmers who were willing to sell a communal tree for a cash payment. Lumber companies quickly cleared hectares of old growth forest, using the Colima Railway to transport the wood. The railway has since stopped running as a result of the decrease of lumber extraction. When the Reserve was declared a protected area, communities were allowed (if not encouraged) to create communal, sustainable lumber extraction companies (Graf et al. 2003; Jardel 1992). In El Terrero, for example, the community relies on sustainable extraction of lumber which they also process at a local timber mill. The local lumber company is the backbone of their economy (supplemented by livestock tenure and ecotourism).

Throughout the lower altitude regions of the Reserve there are series of rivers that flow from the main artery, the Ayuguila River. Fishing for auto-consumption decreased during the 1990s due to heavy pollution of this river system by regional sugar industries in El Grullo and the local cities of El Grullo and Autlán (Graf et al. 2003; Jardel 1992).

Meanwhile, the past twenty years have seen an overall increase in governmental welfare programs. These have been noted to create a dependency on the government by many in this Sierra (Graf et al. 2003). Programs that supplement household income for sending their youth to schools and the availability of government grants for innovations in ecotourism are examples. "Oportunidades" is one specific program having a noted presence in the Reserve, as I discuss in chapter 5.

No single program is unilaterally beneficial to community members and perceptions of the benefits and drawbacks vary greatly. They create economic opportunities and barriers. They engage the local people in a particular mindset of simultaneously being owed by the government for historical oppression, sense of paternalism with the national government, and pride in being the owners and keepers of this natural gem.[2]

Local Perceptions and Governance

The founding of the Reserve was a "good" thing, according to many of my informants. Why? Because there were logging companies looting the forests and there were internal divisions and conflicts among community members. Founding the Reserve as a protected area put a stop, for the most part, to much of the contention surrounding the use of forest resources by outside industries. However, there are still other concerns and conflicts surrounding the Reserve as a protected area.

One primary area of concern is with the Management of the Reserve. Another is with the local university, CUCSUR, in particular the research carried out through the IMECBIO institute. A third area of contention is with other companies or outside industries and foreign investigators.

Problems with the Management of the Reserve have two primary forms: (1) the restrictions of land use and promotion of some sustainable development programs; and (2) the individual field workers and officers of the Management of the Reserve. People often comment that "they don't allow us to farm the land," insinuating a diminished capacity to benefit from their natural resources. However, these concerns are not as surprising as the constant moaning regarding some specific individuals who work for the Management of the Reserve. A former Director of the Reserve was one target for the complaints and rumors of community members. It was hard to even locate someone to fill the position of Director of the Reserve because of internal conflicts which cause the prior director to leave abruptly. However, upon finding the right person for the job, problems and rumors began to spread. The new director was said to be a "racist" and to "not care about the people." Additionally, the Management of the Reserve is reportedly "corrupt, taking bribes from people growing pot [marijuana]." Most of these perceptions come from Ayotitlán; the Management of the Reserve is not nearly as active nor predominant in Zacualpan.

Field workers of the Management of the Reserve are usually trained in agricultural or environmental sciences and appreciate the cultural and ecological value of the Reserve. They recite their job description, which is "to protect the rights of the communities and support them." Furthermore, one field worker commented that "the only ones looking out for the people [of the Reserve] and helping them are the Management of the Reserve, whose sole purpose is to support the development within communities." This sentiment is recited on a daily basis, and no doubt these field workers truly believe that they are supporting the communities in proper sustainable development.

However, the university researchers from IMECBIO are also subjects of many criticisms and skepticism from the community members in the Reserve—again, mostly in Ayotitlán because they have a very limited presence in Zacualpan. The local newspapers (both for Aútlan and CUCSUR) have reported on how upset the people in the Reserve are at the Management of the Reserve, UdeG, and the Jalisco government. The main point of contention is Las Joyas—a scientific research station located within the nucleus zone (where the strictest limitations are in effect). Las Joyas was given to the CUCSUR as a gift by the governor of Jalisco. The use of Las Joyas for student learning, as a field school site, bothers many locals in the Reserve who see it as hypocritical. The most precious part of their territory has been given as a gift by the governor, and is being used as a laboratory for some young college students. At the same time, the CUCSUR has explained that their education paradigm is an integrated approach to community-based research. They assert that the research underway benefits the communities and helps promote conservation.

On occasion, I probed my key informants if it would be better to not have the Reserve as a protected area. This was a hard question to answer, and everyone seemed to have mixed feelings. All of my informants could cite the same reasons for founding the Reserve: "there were (internal) conflicts regarding lumber extraction that—along with discovery of teosinte—lead to forming the Reserve." In this general sense, the Reserve was viewed positively. The Reserve serves the purpose of bringing order and control of natural resources. Everyone in the Reserve agrees, unanimously, that the natural resources and forests there should be conserved—in some way "protected." However, once we get to talking about how that management is occurring, differences of opinion (strong opinions) emerge. I wondered about other models of forest and resource management in Mexico—like the Huichol have or the Lacandon areas in Chiapas. "Well, it is different here. But we are similar because we are also ejidos," I was told. Regardless, I found this observation significant in light of the close relationship with the EZLN. Affiliation with the EZLN is a source of pride in their ethnic identity and gives a sense of being connected with the larger cause in Mexico: indigenous rights and drastic land reform.

There are different ways that large corporations, usually owned by business people in other countries, impact the Reserve. There are "burdens large international corporations are putting on local people and lands." Water and resource use is commonly cited. Big companies pollute and usually do not follow traditional water sharing protocols—rotation of the irrigation water to all of the fields of ejido members. Instead, with money and power, resources are controlled by these large corporations, causing local farmers to struggle even harder to succeed.

Notes

1. The study, "Ecología y manejo de la milpilla perenne (Zea diploperennis Iltis, Doebley & Guzmán) en la ECLJ," (*Ecology and management of perennial milpilla (Zea diploperennis Iltis, Doebley & Guzmán) in the ECLJ*) is currently underway with Reyes Genoveva Jiménez Gómez as the principal investigator and as a collaboration between the Management of the Reserve and IMECBIO. No publications have been made to date. However, past studies in the Reserve have lead to important publications regarding the system of maize cultivation in this area (Benz 1986).

2. Wilshusen and colleagues present a detailed critique of some of the prevalent protectionist-type arguments regarding resource conservation as opposed to community-lead conservation strategies (2002).

Chapter 4
Market Interaction and Indigenous People

In this chapter, I turn to the topic of research methodologies and summarize some of the trends in quantitative anthropological investigations of traditional ecological knowledge (TEK) and political-economic systems.[1] How do we know what TEK is? How can we study the importance of an abstract and intangible system like a market? I first provide an overview of the quantitative studies of TEK, focusing on the various research methods commonly employed in ethnoecology. Anthropologists use techniques such as freelists, pile sorts, objective tests, and semi-structured interviews with key informants to elicit traditional ecological knowledge. Indices or cultural consensus models can be constructed from the data to identify patterns of distribution of TEK in cultural groups. Next, I discuss some key variables that are associated with the patterns of distribution of TEK, including: gender, age, social class, and mode of subsistence. I contextualize current understandings of the nature and distribution of TEK in the context of political-economic systems. In particular, I pay attention to the methodologies used to study market participation and market integration of indigenous people. Finally, I suggest some possible connections between TEK, economic market integration, and biodiversity conservation.

Markets and Traditional Knowledge

There are good reasons to study the relationships between traditional knowledge and the use of natural resources. Cultural groups have highly elaborated ways of being in and utilizing their local environments and anthropologists continue to be engaged with studying these cultural dynamics. Anthropology has been concerned with relationships between cultures and their environments since Julian Steward (1955) in the twentieth century, and even prior to his cultural ecology studies when cultural anthropologists acknowledged the importance of environments for human groups (for example, see Boas and Stocking 1989).[2] Knowledge about the local environment, or ecological knowledge, is organized, distributed within a cultural group in a patterned manner, and transmitted across generations (Berlin 1992) and is thus a robust area of anthropological inquiries.

Indigenous and rural people have often been at the focus of these studies of human-nature relationships and ecological knowledge. Indigenous technology is often associated with sustainability and lower negative impact on the environment. Globally, indigenous people live in areas that are biodiversity hotspots (Stepp et al. 2005). Indigenous and rural people may comprise a sizable proportion of the population in developing countries, and may live in high levels of poverty (United Nations Development Programme [UNDP] 2011). In addition, indigenous and rural people hold most traditional ecological knowledge (Cox 2000).

However, economic relationships also shape the ways that humans interact with natural environments. One way of thinking about economic and social influences on TEK in rural areas is the notion of "market integration," referring to the types and degrees of participation that indigenous or rural communities have in the political economy of regional and global markets. The term market integration is used more in the sense of a political economy, and not strictly in the literal sense of going to a market to sell, or purchase, goods. The political-economic markets that impact local people include the transactions from trade, loans, sale, or barter with individuals from within the community and beyond, but also the market structures and frameworks at local to global scales within which these transactions occur. One way the impacts of neoliberal globalization on indigenous people are studied, for instance, is by measuring the importance of a cash economy in their daily lives. The importance of studying market integration in relation to TEK is that the increasing interconnectedness of local markets across national boundaries is an unfolding process, accompanied by exchanges of knowledge and value systems, along with social and power relations; the flows of knowledge and resources are not unidirectional. As a result, both human-environment relationships and ecological knowledge may be influenced.

A good deal of the research regarding TEK falls within the field of ethnoecology, which is considered part of ethnobiology. The broader field of ethnobiology is both a scientific and humanistic field, looking at the relationships (both past and present) of humans and biota (Stepp 2005). The field includes the study of economic, cognitive, social, symbolic, psychological spiritual, and ecological influences on ethnobiological knowledge. Likewise, the field of ethnoecology shares this holistic and inclusive approach to understanding the many facets of human-environment relationships. In this chapter, I am interested in anthropological studies that look primarily at economic and cognitive dimensions of TEK. These studies generally ask questions along the lines of: How do people know about, or think about, plants and animals? How do people use plants and animals?

The aim of this chapter is specifically to provide an overview of the quantitative methods that have been used most frequently to investigate TEK in the

Americas. I then look at ways in which quantitative studies of TEK have considered the importance of global political economic systems in relationship to patterns in the nature and distribution of TEK, specifically among indigenous and rural peoples. The particular body of research reviewed is embedded within studies of indigenous community development and management of forest resources, which are discussed further in chapters 3 and 5, respectively. Notwithstanding, I suggest some of the reasons that the study of TEK in conjunction with market integration is pivotal to biodiversity conservation.

Studying Traditional Ecological Knowledge

Quantitative Approaches in Ethnoecology

There are many ways to define TEK (Berkes et al. 1995; Berlin 1992; Romney et al. 1986; Toledo 2002; Zarger 2002); I blend these existing definitions to arrive at my working definition as knowledge that: (1) has a patterned distribution, (2) is about how to interact with the local environment, (3) is shared by members of the same cultural group, and (4) is transmitted across generations. As globalization increases the speed and nature of knowledge exchange, some anthropologists debate whether relatively "recent" knowledge about novel environments (such as urban landscapes, for example) which is also shared, culturally-informed, patterned knowledge about environments and transmitted across generations, will be considered TEK. Thus, anthropologists are finding alternate terminology, such as "local ecological knowledge" to perhaps avoid possible constraints or confusion related to determining what constitutes "traditional" (Garcia-Quijano 2007). For the sake of this review of trends in the literature, I am relying on the gloss-term *TEK* to help narrow my focus on indigenous and/or rural people.

Toledo (2002) has suggested that a holistic approach to studying TEK is imperative. Toledo articulates three important dimensions: *kosmos*, *corpus* and *praxis* (2002). The kosmos are the beliefs, corpus the knowledge, and praxis the practices, which are taken together to constitute TEK (Toledo 2002). Toledo's argument for this encompassing approach to studying TEK is central to the development of a theoretical stance in ethnoecology, which he has defined as "the entire repertory of symbols, concepts, and perceptions on nature, and the set of practical operations through which the material appropriation of nature takes place" (Toledo 2002, 513:513). Ethnoecology is conceived as the study of not simply knowledge about nature, but also the use of natural resources.

Another way that TEK can be studied is as *theoretical knowledge*, the ability to name or list a species; and *practical knowledge*, the ability to identify and use a species (Reyes-Garcia 2001). Thus, when Reyes-Garcia and colleagues approached TEK through both theoretical and practical knowledge sets, they approximated two of the three dimensions that Toledo indicates are imperative for ethnoecological studies.

Most studies of TEK measure theoretical knowledge because of the challenges in accurately assessing practical knowledge, especially for larger studies. Proxy measures for TEK—indicators commonly measured as a means of ascertaining knowledge—frequently include things like the knowledge of uses of wild plants, names of plants and crops, and information about the interactions between wild plants and animals. To measure TEK, there are several quantitative methods commonly used by ethnoecologists:

- Transect surveys,
- Freelisting,
- Multiple-choice tests, and
- Goal-oriented tests such as pile sorts. (Bernard 1995)

Each of these methodologies depends on having a common set of cultural knowledge in order to be effective (Romney et al. 1986).

Many quantitative studies of TEK begin with freelists as a way of producing an inventory of known species or other domain of environmental knowledge. However, it is worth noting that freelists are not entirely exhaustive of a category of knowledge that is large—such as edible plants. Freelists may be restricted to a known domain of natural environment usage or other category, or researchers may begin by eliciting the local categories of usable biota and then use successive freelisting to determine the species that pertain to each category. These structured interviews can use different techniques to elicit information about plant and animal knowledge; some use only naming of species, others use photographs, herbarium voucher specimens, or guided tours through gardens or specific landscapes as stimuli. Typically, researchers start with a freelisting exercise and then proceed to structured interviews with one or more key informants. Once an inventory of a specific category or domain of knowledge is determined, it is often the case that researchers will use additional techniques, such as pile sorts, to learn more about the ways that the categories of knowledge are constructed. Local experts from the cultural group can identify and corroborate the "correct" answer for questions about the name, qualities, and uses of a species as another form of verifying the validity of the recorded information. The "correct" answer is usually taken to be the most common response, or most agreed upon response, within the cultural group (Reyes-Garcia 2001). Furthermore, some of these measures (e.g., pile sorts or other goal-oriented tests) can be used to construct a cultural consensus model of TEK. Some analytical techniques for constructing models of TEK include multidimensional scaling, coefficient analysis, and cluster analysis (Borgatti 1996). A cultural consensus model seeks to identify a set of specialized knowledge that has a patterned distribution in a community and is shared by members accordingly (Romney et al. 1986).

It is important to note that these cognitive techniques are limited because we rarely have a reliable or fixed baseline against which knowledge can be measured. Most studies are cross-sectional, representing a particular snapshot in time, and we only have a few examples of longitudinal panel studies that allow a historical perspective (Reyes-Garcia et al. 2006). Studies of TEK are careful to avoid saying that there is a global "loss" of traditional knowledge, since we know that knowledge is a fluid and dynamic cultural phenomenon. Anthropologists in particular resist romanticized accounts of culture in which there is a perceived "loss" because there has always been a continual process of creating and re-creating knowledge and culture.

Qualitative studies of TEK are based in interpretive ethnographic methods, such as participant observation and interviews with key informants. While there is certainly overlap in the field methods used by researchers interested in exploring TEK, one common difference is in the rhetorical style and analytical goals. Qualitative studies, which are not the focus in this paper, frequently take constructivist and phenomenological approaches to understanding the beliefs, knowledge, and practices related to TEK (for example, see Toledo 2002).

Studies of TEK

Research on TEK throughout Latin America and the Caribbean, in particular, has illuminated some important patterns in the distribution of TEK among members of the same cultural group. Studies on the relationships between traditional knowledge and social factors have focused on various explanations of variation in traditional knowledge, such as gender (Caniago and Siebert 1998; Garro 1986), modernization (Benz et al. 2000; Vandebroek et al. 2004), social status (Garro 1986), and income or subsistence activity (Reyes-Garcia 2001; Godoy et al. 1998; Guest 2002).

Godoy and colleagues (1998) found differences in traditional knowledge between men and women associated with the division of labor in Honduras; Reyes-Garcia (2001) found the same in lowland Bolivia. Garro (1986) and Boster (1986) both elected to avoid the confounding factors of sexual division of labor by sampling only women.

Social status, life experiences, and age have been found to be important components of how much traditional plant knowledge an individual possesses. Garro (1986) studied the determinants of traditional knowledge among healers and non-healers in Pichataro, Mexico. Garro's primary objective was to determine if traditional healers and community members shared the same cognitive models of illness and illness treatments. She used structured interviews and a term-frame substitution task (see Bernard 1995) to sample adult women in the community (Garro 1986). She found that individuals did share the conceptual organization of ethnomedical knowledge, regardless of healer/non-healer status, but that older women, because of individual experiences, were more likely to share more traditional plant knowledge (Garro 1986). Age was also an important factor in the amount of traditional plant knowledge an individual possessed. Garro suggested that age was an effective proxy measure for life experiences. However, it may not always be the case that individuals will have more experience with a domain of knowledge simply because of increased age; specialized knowledge is taught and acquired through informal educational networks (Zarger 2002). Furthermore, cohort effects may be a reason for age patterns in the distribution of cultural knowledge.

In conducting a quantitative comparison of intercultural traditional plant knowledge among healers in lowland (Amazon) and highland (Andes) Bolivia, Vandebroek and colleagues (2004) expected to find a greater degree of traditional plant knowledge in the Amazon because it is a more diverse environment. Vandebroek and research team used surveys, fieldtrips with key informants, and samples to construct semi-structured interviews regarding the identification and uses of plants. Socioeconomic information from a regional census was also used to characterize housing type and language. Their results, again, point to the need to further understand the dynamics of traditional plant knowledge variation. They found a higher degree of traditional plant knowledge in the Andean region—unrelated to biodiversity, modernization, or healer status—which they suggest may be associated with social factors in the community and family (Vandebroek et al. 2004).

Benz and colleagues (1994; 2000) analyzed the relationships between knowledge of plant use and various indicators of community change and modernization. In eight communities in the Sierra of Manantlán Biosphere Reserve, Benz's research team used herbarium voucher specimens collected in the vicinity of each community to elicit individuals' knowledge of useful plants (Benz et al. 2000). They used these data to create various community-level indices of ethnobotanical knowledge, which were in turn used to assess the relative importance of different plants across communities. The communities were ranked based on national census data in terms of marginality. They looked at the percentage of households in a community that had certain housing conditions (dirt floors, electricity, plumbing, sewer or one-room only) along with percent of literacy and primary school completion. Their results showed that adoption of modern services in some, but not all, communities does seem to correlate with a decrease in knowledge of local plants (Benz et al. 2000). However, the community of Telcruz confounded their analysis, as there was no apparent relationship between the community's socioeconomic profile and the aggregate measure of plant knowledge for the community. They have suggested that further investigation into the economic activity profile of households and land tenure may more adequately explain variation in traditional plant knowledge (Benz et al. 2000), a topic which I return to later in this chapter.

Reyes-Garcia and her research colleagues have contributed extensively to the field of traditional knowledge and cultural consensus modeling (Reyes-

Garcia 2001; Reyes-Garcia et al. 2006). Carried out under the banner of the Tsimane' Amazonian Panel Study[3] headed by Godoy and colleagues since 1999, the project has produced various studies of the culture and society of the Tsimane' indigenous people who live in lowland Bolivia. The research team of Godoy, Reyes-Garcia, and colleagues, has looked at ethnobotanical knowledge held by all adults in thirteen villages (Reyes-Garcia et al. 2003). They used freelisting to determine a subset of species that were most commonly listed by the majority of their informants, and this subset of species was used to further assess shared TEK. Their study assessed not only theoretical ethnobotanical knowledge, but also reported practical knowledge and skills. Practical knowledge and skill were approximated by asking participants if they have ever actually used the forest resource in the capacity that it is reportedly used and if so (1) how often and (2) how well they completed the task. Reyes-Garcia (2001) details the construction of questionnaires that assessed theoretical and practical knowledge. Cultural consensus analysis was used to assess the results of the theoretical knowledge of local forest resources and self-reported skills in using them. Their research with the Tsimane' indigenous people in Bolivia seemed to show that not all types of market integration would lead to a so-called loss of TEK. Instead, some types of economic activities that were based in cultural traditions or capitalized on local natural resources seems to support higher retention of TEK (Reyes-Garcia et al. 2007).

Ahmed and colleagues (2010) used similar methods to those of researchers looking at the relationships between TEK and market integration, focusing, however, on knowledge of landscapes instead of knowledge of species. They relied on freelists and subsequent interviews with key informants to identify folk taxonomy (Ahmed et al. 2010). Pile sorts were then completed by study participants and the data were used to calculate two types of individual level scores which represent the folk taxonomy: the Land Use Importance Value was determined by the number of markers (points) assigned to each sample by individual respondents, and the proportion of correctly identified tea samples was the individual's Ecological Knowledge Score (Ahmed et al. 2010). I return to this example later in the section on "Relating TEK to Market Integration."

Understanding Market Participation and Market Integration

Economic changes, particularly increased integration with markets outside the community, are related to processes of modernization (i.e., acquiring features of industrial society, including a formalized educational and political system), or development (i.e., having an economy based on industrial production), but these concepts are not synonymous. Economic change does not necessarily imply changes in TEK (Berger and Luckmann 1966). However, consumer products and services, labor patterns, and worldview are interrelated (Godoy 2001). Exposure to foreign-made consumer products and access to new technologies are examples of the pervasiveness of outside markets in peri-urban and rural Mexico (Gerritsen 1998; Olson 2009). Employment, purchases, and relationships with others within the community are all related to the broader structures and relationships between individuals and formal political institutions. Individuals and households increase their stake in the market economy as they transition from relying on trade and self-sufficiency to reliance on market interactions and products. Across various settings, scholars have used these related concepts of modernization, development, and market integration in very similar manners.

Godoy and his colleagues (1998; 2001) have measured and conceptualized "market participation" in various ways for indigenous, non-industrialized communities. Market participation, or market integration, is a concept that aims to quantify the amount and ways that individuals and households are engaging in regional and global economic transactions. Market participation can be measured by taking the sum of monetary earnings, along with the monetary value of items received by barter or trade, and determining the proportion of cash transactions to the total household income (again, in monetary values [Godoy et al. 2005]). The total household income is the monetary value of all of the things that are consumed or received (including rented or borrowed items) in the household (Godoy 2001). This requires obtaining information on the monetary values of the goods and services received or consumed and also tabulating the money spent by the individuals who comprise the household. Because calculating the measure of market participation in monetary values requires informants and researchers to determine monetary values for things that are consumed in the household, it can be an arduous and error-prone task to calculate this data for non-industrialized indigenous people.

There are other measures which are less accurate, including: measuring the value of commercial and physical assets; quantifying the money borrowed from non-community members; and the distance to village/market town (Godoy et al. 2005; Lu 2007). Market participation can also be inferred from a combination of proximate measures, including geographic accessibility, knowledge of forest resources, and household consumption of market items (Godoy et al. 2005).

Quantitative studies of market participation struggle with a few common shortcomings. The main areas of concern are the tendency for small sample sizes and lack of many rigorous longitudinal or panel studies (Godoy et al. 2005). Most anthropologists and social scientists interested in the phenomenon of market participation have carried out cross-sectional or short-panel studies spanning a one- or two-year period. As with quantitative studies of TEK, there is no consistent system for measuring key variables, making cross-cultural analysis difficult (Godoy et al. 2005). Godoy and his colleagues (2005) also note that researchers need to be careful about plausible reverse causality in market participation studies. Furthermore, individuals may self-select to participate (or not) in markets, which will impact the types of market interactions that are observed within and across cultural groups (Anderson 2002). There are variables that are related to acculturation which need to be sufficiently accounted for, as well, since we know that market participation includes the exchange of knowledge and attitudes, not simply money (Godoy et al. 2005).

Attitudes and values and even language can be impacted by market participation. In the applied social sciences, there have been a series of complementary hypotheses about why people participate in markets (Gonzalez and Nigh 2005; Morgan 1993). Individuals' attraction to, and the appeal of, foreign and novel goods can be a reason for market participation. The encroachment of outsiders into rural and indigenous communities has also been linked with increasing market integration (West et al. 2006), as has the increased presence of national governance or bureaucratic programs (Graeber 2011). The combined impacts of local community population growth and the scarcity of local resources can be reasons to increase market participation, as can the desire to improve individual well-being and socioeconomic status (Alcorn 2001). By contrast, some of the literature has also suggested that there may be distaste for participation in markets and outsiders (sometimes as part of resistance or indigenous-revitalization movements), and these can be reasons that individuals consciously choose to limit market interaction (Anderson 2002; Gerritsen 2004).

Relating TEK to Market Integration

As can been seen, many quantitative studies of TEK have taken into account issues related to economic development and market integration, finding gender, age, class, and socioeconomic status, for instance, to be among important variables (Buenz 2005; Chant 1991; Garro 1986; Godoy et al. 1998; Reyes-Garcia 2001; Vandebroek et al. 2004). However, none of these variables is isolable and can be used alone to explain the variation observed in the distribution of TEK within a cultural group. The effects of integration with market economies are important for the social, economic, and cultural dynamics of indigenous communities, and can have impacts on both TEK and local subsistence strategies.

While Benz and colleagues (2000) found that acculturation might be associated with a loss of TEK in the Sierra of Manantlán Biosphere Reserve of west-central Mexico, much research has shown that there is not a direct relationship between increased modernization and a decrease in traditional knowledge (Benz et al. 2000; Godoy et al. 1998; Guest 2002; Vandebroek et al. 2004). For instance, Buenz (2005) found that language, generational loss of knowledge, and a decrease in biodiversity are all poor indicators of traditional plant knowledge loss in developing countries. Similarly, no change in TEK of children in relation to socioeconomic change was found in southern Mexico in Chiapas (Zarger and Stepp 2004).

By contrast, Ahmed and colleagues' (2010) results showed that there are meaningful changes in values, subsistence strategy, and TEK with an increased interaction with a cash economy. Gender was an important variable in their findings, since not only did individuals' TEK increase with more market participation, but there was a shift from women to men having higher ecological knowledge scores. In this case, an increase in the economic value of tea led to middle-aged men showing increases in the use of ecological knowledge, involvement with tea production, and the number of names and terms of tea (Ahmed et al. 2010). Significantly, they found that there was no single cultural model of tea knowledge because the contents of the specific body of knowledge was dynamic and transitioning to meet the needs of community members and markets.

Ahmed and colleagues' (2010) work resonates with my findings regarding medicinal plant knowledge, which I present in detail in chapter 8. In the Reserve, I found that the two indigenous communities that were closest and furthest from a market town had the least amount of agreement (i.e., intragroup consistency and shared responses) about theoretical TEK; the indigenous community mid-distance to the market town had the most agreement about theoretical TEK (Olson 2009). I found that the three study communities each had distinct sets of medicinal plant knowledge, despite sharing similar cultures, albeit with varying socioeconomic community profiles. The community that was the most geographically isolated had higher rates of internal agreement about the TEK I measured than did the other two study communities.

In a similar vein, in Amazonia and Honduras, Godoy found that individuals can have higher amounts of TEK if that knowledge is directly related to a market interest (Godoy et al. 1998; Godoy 2001). Godoy and colleagues studied the effects of markets on Tawahka Indians in Honduras. Individuals who sold more rice had less knowledge of forest products, while those who sold forest goods had more knowledge of plants (but not animals). Taking form of livelihood into account, they found that those who sell a particular type of wood which has a high market value, in fact had higher knowledge of both plants and animals in the forest. Thus, they found that the type of economic activity correlated with the amount of traditional forest usage knowledge an individual had: those who sold agricultural products had less traditional knowledge of uses of forest products than those who sold forest products at market (1998).

In lowland Amazonia, Reyes-Garcia (2001) found that TEK became more specialized among individuals who worked more extensively in, and with, the market economy. Guest (2002) studied Ecuadorian fisheries and found that participation in a market economy could spur the development of new areas of TEK driven by market opportunities.

In sum, studies of changes in TEK in relation to market integration have shown a wide variety of correlations. On the one hand, several studies have found market integration is related to the creation of new ecological knowledge. This could be because of competitive advantages gained in relation to market opportunities (Grant 1996). Indeed, some research has confirmed a trend to increased types of specialized TEK with an increase in market interaction (Guest 2002; Godoy 2001; Ahmed et al. 2010). Yet there have been many studies to the contrary, too, which seem to show a decrease in TEK when market participation increases (Nolan and Robbins 2000; Reyes-Garcia 2001; Srithi et al. 2009; Zent 2001; Benz et al. 2007). Yet even other studies have found a neutral or no apparent relationship between market integration and TEK (Ayantunde et al. 2008; Zarger and Stepp 2004).

TEK, Market Integration, and Biodiversity Conservation

There are some implications of quantitative ethnoecology research for conservation of biodiversity. There are good reasons to think that economic development of indigenous people will be related to shifts in natural resource use and shifts in knowledge about natural resources (Zarger 2002). From a historical perspective, increased market participation and population growth has tended to lead to degradation of renewable natural resources because of increased pressure on the local environment (Alcorn 2001). Moreover, the notion of an "ecologically noble savage" (Redford 1991) is, in any case, flawed because traditional subsistence styles have varying degrees of impacts on natural resources, regardless of market participation.

Increased market participation means likely changes in the ways that local natural resources are used, in addition to associated changes in the diversity of crops grown. Some researchers have found that the sale of non-timber forest resources and products in markets can serve to enhance conservation because of the associated market value (Ahmed et al. 2010; Godoy 2001). By contrast, the production of mono-crops in tropical forests for sale to non-local markets leads to deforestation by virtue of many possible factors, including: pressures to increase farm productivity in peripheral areas, pressures to increase the size of the area under production, or the introduction of crops that cause environmental degradation. With shifts in the usage of local natural resources, we can expect a ripple effect on the associated bodies of ecological knowledge that may be informing resource management strategies.

Such changes can have varying implications for both TEK and conservation. For instance, the increase in the folk taxonomy and names of tea varieties that Ahmed and colleagues (2010) found corresponds to what Berlin predicted about an increase in human use of a natural resource leading to an increase in the affiliated folk taxonomy (Berlin 1992). Possible long-term effects of market integration may include perceptions that some aspects of the social and cultural fabric of a community are deteriorating (Li 2007). In addition, emigration from traditional landscapes may be associated with a reduction in TEK held by community members (Rigg 2006).

Studies of TEK have frequently focused on non-industrialized indigenous people in rural areas. The communities where TEK has been studied are sometimes also in natural protected areas, which is an important element in the content, distribution, and nature of TEK. As I explained in chapter 1, protected areas receive special protection (normally through laws and policies) because of their unique environmental and/or cultural value. Natural protected areas have natural resource management regimes that may or may not be informed by cultural knowledge, yet constrain the ways that local people interact with their environment, thereby potentially affecting local knowledge of how to identify and use different species (Zarger 2002).

Agroecosystems, with their potential to reconcile biodiversity conservation with agricultural production, may have other implications for TEK (Ahmed et al. 2010; Alcorn 2001). Ahmed and colleagues (2010) carried out a study of upland communities where indigenous and low-intensity tea agroforests are still used in the Yunnan region of Southern China. Their research was interested in the role of market integration on TEK and an indigenous agroecosystem. As described earlier, their measures indicated an increase in the amount of ecological knowledge held by individuals who become more engaged with the tea market (Ahmed et al. 2010). The findings from Ahmed and colleagues (2010) were also significant in a number of other regards. There was an overall increase in the total quantity of important land use categories. This is related to a diversification of subsistence activities and an increase in the perceived labor investment in different types of cultivation. Additionally, due to the increase in the market value of "ecological" tea, tending the tea agroforests became an enterprise for the entire community whereas in the past this crop had been tended by women. Along with this came the creation of new roles as market intermediaries for a set of local men. The impact on the household subsistence strategy was marked—and tea agroforestry was now the key cash income for most households. The income from tea agroforestry was associated with an increase in modern household items such as cars, TVs, clothes, meat, snack foods, and so on. The shifts in material consumption were also accompanied by shifts in social capital and the construction of roads, schools, and other institutions which help transmit ideologies and knowledge (Ahmed et al. 2010).

Although the studies reviewed do not show a linear relationship between increased integration with outside markets and a loss of TEK, increased understanding of this relationship, as well as processes that are inclusive of indigenous (Berlin and Berlin 2004; West et al. 2006) and rural people and their TEK, offer opportunities for creative approaches to conservation and development (Anderson 2002). For example, some strategies for integrated conservation and development (ICDP, as discussed in chapters 3 and 6) have seen opportunities to use cash payments to avoid overexploitation of forest resources. The stipends are intended to supplement the household income in exchange for non-use or restricted use of forest resources. However, the cash payments may be seen as a paternalistic model that creates dependency on the government instead of community empowerment. Furthermore, it is possible that cash payments merely serve to shift the extraction of forest resources, and intensive production of non-forest products, to other environments. Agroforestry has been suggested as an intermediary strategy that may be able to maximize local economic and social benefits while avoiding environmental degradation.

Future Directions

Quantitative studies of TEK have looked at a variety of relationships, such as: the well-being of indigenous peoples; the role of participation in the market economy; and the use of renewable forest and other natural resources. However, quantitative studies struggle to find a consistent form of measurement that would enable developing broader hypotheses and theories about these relationships (for more on this, see Godoy et al. 2005). While anthropologists have been essential in working on the question of what roles market participation plays in the transmission and creation of TEK, and in documenting TEK, Cox (2000) points out that studies of ethnobotany have also been an important area of inquiry for other fields, such as pharmacology and conservation biology. Interdisciplinary and multifocal research may have the potential to illuminate more of the ways that the creation, distribution and transmission of TEK are related to the ever-growing prevalence of outside markets in small, rural, indigenous communities. Investigations about traditional ecological knowledge will need to continue to be attentive to equity issues related to the protection of indigenous peoples' intellectual property (Berlin and Berlin 2004).

Forest and other natural resources have multiple kinds of value, and economic factors, particularly markets, play important roles in the arena of conservation and natural resource management. There are both theoretical and practical benefits of studying the topic of TEK and market participation. Theoretically, understanding the complexities of TEK, particularly its patterned distribution, advances our understandings of the ways that humans interact with diverse environments and cope with novelties in environments. Practically, there is great interest in managing the human impact on our environments (and the impact of our environments on us). Anthropologists have advanced hypotheses about the ways that sociocultural change unfolds in relation to globalization and increasing market integration. Testing these hypotheses informs the ongoing programs associated with conservation and sustainable development.

There are direct implications of this research for conservation. Parks and natural resource protected areas are a hotly discussed model for conservation (West et al. 2006). In the global North (or Western countries), conservation of natural resources is being taken-up in a neoliberal capitalist model which directly impacts conservation in the global South. The commoditization of natural resource knowledge in conservation and sustainable development programs is related to these capitalist models of resource use, and to assumptions that forest knowledge will be reified when it has economic (market) value. For these reasons, anthropologists continue to study the relationships between economic activities and traditional ecological knowledge, to date finding both positive and detrimental interactions.

Notes

1. This chapter is based on my review article, "Anthropology and Traditional Ecological Knowledge: A Summary of Quantitative Approaches to Traditional Knowledge, Market Participation, and Conservation" which was published in 2013 in *Culture, Agriculture, Food and Environment* 35(2):140-151.

2. It is worth noting that Franz Boas was an early American Anthropologist who, by way of his research with American Indian tribes, took frequent note of the indigenous knowledge of environments and wildlife (as republished in 1989). Other early anthropological accounts of traditional ecological knowledge were done by Frank Hamilton Cushing (for example, see Cushing and Green 1979). For additional discussion of the origins of the term "ethnobotany" see Harshbarger (1896) and the later work of J. P. Harrington. For additional information about the history of anthropologists working in ethnobotany, see the article by Ford (1985).

3. For more information about the Tsimane' Amazonian Panel Study visit http://tsimane.org/.

Chapter 5
Governance: Community Decision-Making
With Co-Author Peter R. W. Gerritsen

Sustainable development and conservation projects are inherently political, multi-axial and interdisciplinary. In fact, the most difficult aspect of successfully realizing sustainable development projects is bridging the political and ideological differences between stakeholders (Devlin and Yap 2008; Wilshusen et al. 2002). This chapter addresses the processes of sustainable development and conservation and local perceptions of actors in the Reserve where (within the same cultural group and resource management program) there are different community dynamics that have profound impacts on sustainable development processes.[1]

A number of key stakeholders involved in projects in the Reserve are outsiders to its communities. Government agencies, research groups, and private industry can all be characterized as secondary stakeholders. From a local perspective, they are "outsiders" since they are not only foreign to the community, but are often perceived as having ideologies different from or even contrary to those of community members. In the process of working to implement sustainable development and conservation programming, local-level decision-making is considered critical to successful outcomes (Chambers 1997; Wallerstein 2006; Wilshusen et al. 2002). Consequently, the formation of allegiances, intragroup conflicts, political affiliations, relationships with scientific investigators, and major businesses all contribute to the ability—and inability—of groups to reach consensus and realize sustainable development activities (Graf et al. 2003; Berlin and Berlin 2004; Rosenthal 2006).

The multiple relationships between community members, their organizations, and outside change agents have significant impacts on the outcomes of conservation and sustainable development (Graf et al. 2003; Jardel 1992; Gerritsen 2002). I want to focus now on the local-level relationships and decision-making structures in the Reserve. In this chapter, I am going to illustrate some of the nuances in decision-making within the same protected area and cultural group, which are essential to community-based resource management by using Ayotitlán and Zacualpan as examples. Recall that Telcruz is part of the same ejido as Ayotitlán.

Critiques of development have emphasized the need to move away from Western ideals (Escobar 1995). In the move toward community-based approaches, we find that in many ways the goal outcomes are unchanged from the capitalist model for development that was dominant during the 1980s and 1990s (Ferguson 1994). Social scientists have found that the same underlying paradigm of sustainable development dominates in many community-based strategies (Pfeiffer and Nichter 2008; Brechin et al. 2002). The case studies that follow illustrate some important components of community-based sustainable development within the context of protected natural areas. Because community-based development is a process, community dynamics are significant factors in programming. The dominant paradigm for sustainable development in Mexico is specifically geared toward the integration of community members in resource management (Halffter 1995).

Sustainable development aims to assure the immediate, long-term, and future benefits of human and natural resources to communities (Anand and Sen 2000). In Mexico, the explicit goals have also been to empower indigenous and underprivileged communities to improve their economic and health statuses through the managed exploitation of their human and material resources (Martinez et al. 2006).

Community capacity-building means integrating existing local strengths with research, new knowledge, skills, and resources to increase the power of community members (Wallerstein and Bernstein 1994). Wallerstein (2006) has argued for integrating empowerment and capacity-building within development projects.

As sustainable development and conservation continue to be important goals to achieve in the Reserve, building capacities within communities and respecting autonomy are important aspects of management processes (INE 2000). In particular, the local institutions in charge of natural resource management are pivotal in the success of ICDP.

Participation and Observation at Community Meetings

During my research periods in Ayotitlán and Zacualpan, I participated in two governance meetings in Zacualpan and six governance meetings in Ayotitlán.[2] I provide some vignettes from my field experiences to highlight processes of decision-making, considering these to be characteristic of the study communities. These two communities, Ayotitlán and Zacualpan, have starkly contrasting socioeconomic profiles; Zacualpan has reached generally higher socioeconomic levels than Ayotitlán (INEGI 2005). Both of the communities are governed by the same mandate for the Reserve, but there are marked differences in the socioeconomic profiles and access to outside markets and consumer goods. From

March through August 2007, I made various community presentations regarding my ongoing anthropological research at community meetings. Throughout these meetings, I recorded copious field notes. Of course, the proceedings of the governance meetings only provide a portion of the context—and the processes of decision-making spill into the everyday lives of individuals. Unstructured interviews with key informants were also conducted. We have also supplemented the analysis with relevant literature on processes of decision-making in regards to sustainable development and conservation in Mexico.

The specific comparison of the contrasting socioeconomic profiles presented in the next section establishes a general image of each community. Following that, the analysis of the two study communities is structured by common variables emphasized in many ICDPs: (1) social characteristics, (2) economic characteristics, (3) political characteristics, and (4) conservation characteristics.

Community Structures and Profiles

As explained in chapter 2, there are two primary types of community land tenure in Mexico, and which are present in the Reserve: (1) *comunidades indígenas* (indigenous communities) and (2) *ejido. Comunidades indígenas* are a traditional land tenure system wherein local people claim an ancestral connection to a specific territory. In its contemporary form, an *ejido* is a combination of the Spanish and Nahua tenure systems. The ejido is the legal entity with elected officials responsible for managing communal grazing lands and external affairs. The president, secretary, justice of the peace, and council of members are common positions within the governance structure in all ejidos, but some also designate a treasurer.

Within the Reserve there are five indigenous communities and thirty ejidos. Zacualpan is an indigenous community with a governing body (called the Commission of Communal Goods) consisting of a president, secretary, and justice of the peace. Though Zacualpan is situated on the border of the Reserve, it is under the jurisdiction of the Management of the Reserve because almost all of the land held by this community is within the limits of the Biosphere Reserve.

Notwithstanding its indigenous origins, the original land tenure regime in Ayotitlán was not reconstituted as an indigenous community, but as an ejido during the mid-twentieth century (Rojas et al. 1996). Secondary stakeholders bore a strong political influence on the decision to constitute Ayotitlán as an ejido. The establishment of an ejido opened up the unlimited exploitation of rich forest resources that were present within the limits of Ayotitlán by these same secondary stakeholders (Brockmann 1994; Rojas et al. 1996).

As part of a movement for indigenous revitalization in the late 1990s, the *Consejo de Mayores* (Council of Elders) was reestablished. The main role of the Consejo de Mayores is a cultural advisory one, due to the fact that it is not officially recognized by the Mexican state. Formally, the Commission of Communal Goods (comprised of a president, secretary, justice of the peace, and council of members) is the legal governing body in Ayotitlán. Ideally, these two institutions work together in advocating for community concerns and needs in Ayotitlán. The indigenous community of Zacualpan does not have a formal Council of Elders, but does have a Commission of Communal Goods that is the elected, legal, governing entity.

Recent census data indicates that there is variation within the Reserve at the community level using (Benz et al. 2000; INEGI 2005). The range from "traditional" to "modern" consumption patterns and socioeconomic status is present in different forms. This is illustrated most dramatically in the communities of Ayotitlán and Zacualpan. The total number of private households in Zacualpan (390) is more than two times that of Ayotitlán (142); Zacualpan is notably larger and peri-urban, whereas Ayotitlán is small and rural by comparison. As shown in table 5.1, most houses in both Zacualpan and Ayotitlán have more than one bedroom. However, the differences in use of modern amenities are much greater between the two communities; while in Zacualpan most have a refrigerator (63 percent), in Ayotitlán very few do (20 percent). Other modern conveniences like washing machines are little represented in Ayotitlán (4 percent), although Zacualpan again boasts significantly more (20 percent). Illiteracy in individuals over fifteen years of age is three times greater in Ayotitlán (15 percent) than Zacualpan (5 percent). These general differences in community characteristics reflect the degree of "modernization" and "tradition" in the two communities.

Table 5.1. Comparison of households in Zacualpan and Ayotitlán, select characteristics.

	Zacualpan	Ayotitlán
Inhabited private residences with dirt floors	13.6%	42.3%
Inhabited private residences with one bedroom	37.2%	39.4%
Inhabited private residences w/ plumbing in bathroom	87.9%	20.4%
Inhabited private residences without public water	3.6%	53.6%
Inhabited private residences with electricity	96.9%	85.2%
Inhabited private residences with television	85.6%	31.7%
Inhabited private residences with a refrigerator	62.6%	20.4%
Inhabited private residences with a washing machine	20.3%	3.5%
Population 15 years and older that is illiterate	4.5%	14.9%
Total inhabited private residences	*390*	*142*
Total population	1724	683

Source: Instituto Nacional de Estadística y Geografía (INEGI). "Iter2005." Sistemas Nacionales Estadístico y de Información Geográfica, 2005.

The presence of a relatively more cash-income-oriented economy is evident in Zacualpan by the more urban-style community center, the prevalence of cars, and the cement-metal construction of houses. In Zacualpan, less than 14 percent of houses have dirt floors, and most (88 percent) have bathrooms equipped with public sewage pipes (INEGI 2005). In this indigenous community, more than 85 percent of households have television and virtually all have electricity (INEGI 2005). These material manifestations of access to paid labor (albeit typically as low level "peons") also demonstrate increasing interactions with outside culture, urbanity, and education. Waged labor jobs are found in factories or agricultural greenhouses in nearby Colima.

Agriculture and livestock tenure are characteristic economic activities in the Reserve. Whereas in Ayotitlán, small-scale farming with low levels of livestock tenure is common, in Zacualpan, waged labor and livestock tenure are more prominent. Social welfare programs are a very important component of the household economic stability in the Reserve, and also a very prominent social issue.

The Sierra of Manantlán region is notorious for its history of conflict over forest resource management (Gerritsen 2002). Community members traditionally exploited natural resources. However, the traditional means of management in Ayotitlán became corrupted as people became influenced by aspirations of power and money and outsiders came to buy forest resources (Rojas et al. 1996). The current need to balance conservation and development goals contributes to the complexity of decision-making at a community level in the Reserve (Brechin et al. 2002). In recent years, national social movements that are part of a broader struggle for agrarian reforms and indigenous rights in Mexico have helped draw political attention to the area.[3]

Members of two key outside stakeholder groups (i.e., representatives from the local university and the governmental management of the Reserve) have attributed the marginalized status of Ayotitlán to its dependency on government-sponsored welfare, referring to the Mexican program *"Oportunidades"* (Opportunities), a program that evolved out of the 1997 program *"Progresa"* (Progress). This program provides a monthly stipend for all children under the age of five and a monthly stipend for all school-aged children enrolled in and attending school. This program is prominent in Zacualpan, as well, but outsiders have not cited it as a factor in their socioeconomic status.

People in Ayotitlán, at the same time, feel frustrated. They are prohibited from exploiting their natural resources—the remaining forests are located in the restricted core zone of the Reserve, and they perceive that others are benefiting from their land and forests. The two case studies presented in the following sections highlight some decision-making processes in these two communities. The regional and national procedures that pertain to sustainable development and conservation in the Reserve are realized in very different ways across the two communities; the outcomes of ICDP projects are different for both Zacualpan and Ayotitlán. The case studies point us to new ways of appreciating community dynamics in rural areas in general, and in protected natural areas, in particular.

Description of Outside Intervention in the Reserve

For the purposes of this chapter, stakeholders and key actors are classified in four groups: (1) community members and institutions, (2) state-sponsored conservation and development agencies, (3) scientific research institutions, and (4) for-profit companies. These four types of key stakeholder groups have different relationships with the communities and the other stakeholder groups. As schematically indicated in figure 5.1, the types of relationships can vary by both the nature and degree of interaction.

Community leadership and institutions are integral components of the dynamic process of representation and decision-making in the two study communities. Naturally, there will be variation in the opinions and objectives within a community, and for this reason it is important to consider the relationships between representatives of the community and of the stakeholder groups. Furthermore, the accountability of leaders to their constituents is an essential part of the decision-making process. Scientific research often has goals that do not always prioritize benefit to the communities and can thus be perceived as "draining" the community of resources and value. Businesses and industries are likewise often seen as groups whose extraction from the communities outweighs the benefits received by community members.

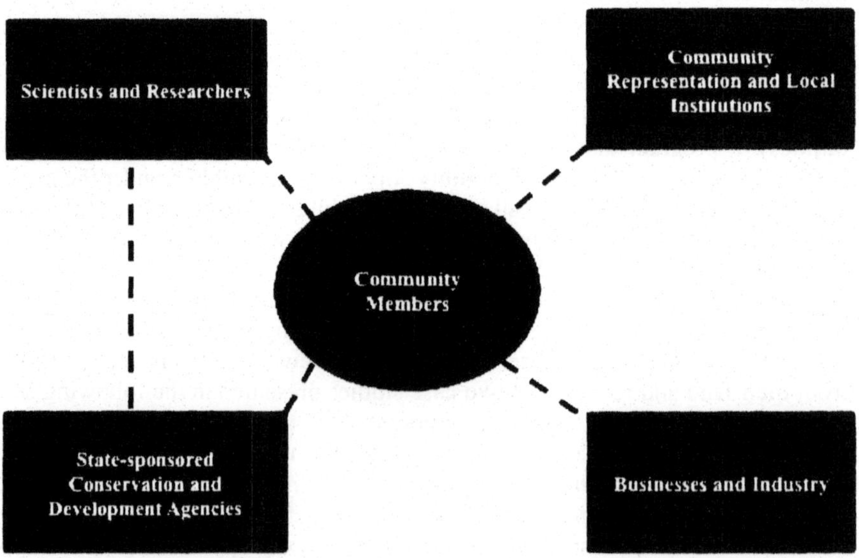

Figure 5.1. Dominant relationships between key stakeholders in community-based participatory research projects.

As depicted in figure 5.1, key stakeholder groups from outside the study communities include the governmental management of the Reserve and the researchers at the University of Guadalajara. Both of these two entities have interests in the sustainable development programs of the Reserve.

The expressed goals of two outside stakeholder institutions are complementary to the goals of ICDP. The IMECBIO (from the Spanish *Instituto Manantlán de Ecología y Conservación de la Biodiversidad*) is an institute created in 1985 by and operating within the University of Guadalajara. At IMECBIO the pedagogy centers on scientific research and its application within the research communities (Martinez et al. 2006). Scientific research in the Reserve pertains mainly to the biodiversity of plants and crops. The botanical information is collected in the agricultural fields and forests belonging to the respective community members.

While many of the actual field researchers are students from the communities who study at the University of Guadalajara, the lead investigators tend not to be indigenous people, but outsiders who have come from urban areas. Students from the Reserve who study at the university are enrolled in academic programs related to the study and management of natural resources. Said students are able to conduct research within the Reserve without going through the community consent process. Meaning, they are allowed to carry out surveys and collect plant specimens without obtaining any special permits. Having these students on research teams may help facilitate positive community relationships because the other community members can more easily identify with the research project if they perceive a personal connection. Additionally, these students may bring a specialized expertise and knowledge of the geographical area that contributes to the research being conducted.

Due to the technical expertise and resources of the IMECBIO, the academic research institute is an important component of the sustainable development process in the Reserve (Martinez et al. 2006). The governmental management of the Reserve and the IMECBIO share the same conservational and sustainable development goals. The communities within the Reserve are hubs for interaction between these two stakeholder institutions. In general, the IMECBIO provides research that the Management of the Reserve then applies to its administration of the Reserve.

A variety of research and developmental activities are undertaken in the Reserve, and the success of these activities is measured in different ways. At IMECBIO the production of scientific data is the key output, while at the Management of the Reserve the realization of community projects is the primary measure. Ecological studies are predominately used to monitor and track biodiversity and thus have a direct application to conservation programs. The implementation of sustainable development projects is predominately the work of the Management of the Reserve, where the focus is on procuring and maintaining monetary investment from grants or other agencies, in addition to some qualitative assessments of community engagement. Community capacity building is seen in the continuation and growth of projects as well as in the tangible benefits demonstrated in improved socioeconomic conditions and access to resources.

The impact of scientific research is less direct because it is intangible (Pfeiffer and Nichter 2008; Wallerstein 2006).

Community Dynamics in Zacualpan

The community members of Zacualpan are characterized as more "modern" in terms of their lifestyle and ideologies. Locally, decision-making is streamlined through the existence of a single elected body of representatives, which is the Commission of Communal Goods (president, secretary, justice of the peace). The recent completion of a sustainable development project is an example of the apparent success of community-based resource management in Zacualpan.

Zacualpan sits on a river near the large urban area of Colima. To capitalize on this fortuitous location, and help promote environmental stewardship, a day park was created on the river bank. The park includes picnic areas, marked trails to encourage controlled use of the river's edge, and a small eatery. The Management of the Reserve developed the plans for the park. The actual process of creating the park—from concept to the grand opening—lasted more than eighteen months. A community-based advisory board for the project was created in Zacualpan in the form of a special committee on tourism that was comprised of local community members.

The primary source of funding for the river-side park project was procured by the Management of the Research from another national agency, the Comisión para el Desarrollo de Comunidades Indígenas (Commission for the Development of Indigenous Communities, or *CDI*). The CDI works directly with indigenous communities in Mexico to design and carry out a variety of sustainable development programs. The CDI operates parallel other key stakeholders, namely the Management of the Reserve. Whereas the Reserve operates beneath the SEMARNAT, the CDI is under the national jurisdiction (meaning the Mexican Congress and President). Again, this means that there is a need for communication and collaboration between stakeholders in order to maximize the efficient use of resources.

Zacualpan has demonstrated an active interest in the programs and initiatives of outsiders. Their receptiveness to interactions with outsiders is, in part, associated with their increased levels of interaction with urban areas and the outside world via television, radio, and other more "modern" mediums. Their proximity to Colima, approximately thirty minutes by public transportation, can be attributed as a key factor of this. In fact, their local health clinic is minimally staffed and serviced since people go to Colima for their medical needs. Most community members hold to their shared identity as an indigenous community while working as paid laborers in nearby Colima. In comparison to the average socioeconomic status in Colima, individuals from Zacualpan tend to be poorer (INEGI 2005). Their expressed interest in sustainable development programs (like ecotourism), the simpler political structure which is easier to navigate, and the higher overall level of education in Zacualpan, have contributed to the viewpoint that the community is "successful" by representatives of the Management of the Reserve and other key institutions from the government or academic sectors.

Community Dynamics in Ayotitlán

Ayotitlán presents a contrasting socioeconomic profile and set of interactions with outside stakeholder groups to those of Zacualpan. The community is characterized as more "traditional" and rural, with an ideology that is often less in-line with that of the outsiders with whom they interact. Furthermore, the history of Ayotitlán in regards to projects and sustainable development interventions has been marked by more violence. Consequently, community members are skeptical of the resource management program and rely on a system of feedback and consultation with local leadership structures in order to guide decision-making (as illustrated in figure 5.1). The process of decision-making in Ayotitlán is not linear because of the parallel leadership bodies. Taken together, these factors may create the impression to some members of stakeholder groups that there is a problem or barrier to community-based resource management.

Outside researchers and scientists are expected to go through the process of obtaining prior consent from the community to carry out research—whether in Zacualpan or Ayotitlán. However, in our experiences, the actual process is only formally adhered to in Ayotitlán. This component of the research and programming process speaks to issues of legitimacy of the process of implementing an ICDP (Brechin et al. 2002).

In Ayotitlán, the Council of Elders speaks up for a community whose members are otherwise very quiet during assemblies and meetings. It is the moral and social responsibility of the Council of Elders to be the voice of representation in the community. Yet some members of outside stakeholder groups can perceive the existence and persistence of this group as redundant. The members of the Council of Elders do question the goals of outside researchers; they ask what the primary activities will be and how the community members can expect to benefit from the research activities. Additionally, they continue to demand copies of all research work and publications resulting from research carried out within Ayotitlán. In asserting the proper, legitimate process for interacting with outsider stakeholder groups, the community of Ayotitlán demonstrates strong organizational and institutional capacities.

Having a designated group for moral and social leadership can be useful in asserting opinions that are contrary to those held by members of outside stakeholder groups. The Council of Elders has both a symbolic and functional role in promoting the value of local culture and community ideals. However, for many outsiders this mainly political function is downgraded to a cultural issue, as it seriously undermines outsiders' room to maneuver in the community.

As part of a national struggle for agrarian reform and indigenous rights originating in the Zapatista uprising in 1994, there is an ongoing need for social action in regards to forest resource management. The outcome of a Zapatista congress held in Ayotitlán in 2003 was the signing of a declaration by all of the communities in the Reserve (Delegados de Los Pueblos Indígenas 2003). The declaration, in very strong (political) language and an indignant tone, asserts the rights of the Nahua people of the Reserve and condemns many of the outside stakeholder groups who have in some way been involved in projects in the Reserve but without retribution to local social development. In this declaration, the Management of the Reserve is (erroneously) rebuked for their role in the exploitation of indigenous people and for allegedly helping an outside company steal intellectual property (the specific incidence mentioned appears to have been an incomplete understanding of what occurred). Assembled in Ayotitlán, the leaders of the Zapatista social movement and the communities of the Reserve used the Declaration to reject the constant stream of outsiders in their community forests and lands. In the Declaration, the local leadership insist on being involved in the planning and execution of any conservation and development projects (Delegados de Los Pueblos Indígenas 2003). Additionally, the Declaration stipulates that the products of the research conducted in the Reserve needs to be presented to the communities where the research is carried out.

In either of the two study communities we have described, it is possible to form ad hoc associations for specific purposes, but such associations have only been formed in Ayotitlán. This is another way in which the decision-making process involves more layers, groups, interests and factions in Ayotitlán. These independent associations represent the economic and resource management goals of their constituents and can often be a very demonstrative way of forming special interest parties.

Comparison of Local Dynamics

Thus far, I have sketched an ethnography of community dynamics related to sustainable development decision-making in two rural communities in the Reserve. Now, I will summarize and relate the descriptions of decision-making to some theoretical perspectives on ICDP, which are presented in chapter 6. The two case studies give an overview of local-level decision-making in the context of sustainable development on a protected natural area. In the two case studies, we emphasized some of the critical ways that decision-making structures operate, and provided a comparison of the different sets of community dynamics. The perceptions and attitudes of outside stakeholders are essential to understanding local perceptions and processes related to community-based natural resource management. Throughout the processes of decision-making, the interactions between stakeholder groups are loaded with these implicit and explicit viewpoints.

The following discussion is structured around five broadly construed characteristics of each community including the general, social, economic, political, and conservation attitudes and trends we have provided in the case studies. Table 5.2 highlights some of these factors in the two communities and can help further illustrate the origins of divergent perceptions of decision-making processes.

Table 5.2. Summary comparison of Zacualpan and Ayotitlán attributes.

Variable	*Zacualpan*	*Ayotitlán*
General	Indigenous Community Close to Colima	Ejido Distant from urban areas
Social	"Modern" Accepting of "outsiders"	"Traditional" Skeptical of "outsiders"
Economic	Paid labor (as "peons") Low level of participation in state-sponsored assistance programs. Outside industry is considered acceptable.	Work the land ("campesinos") State-sponsored assistance programs are prevalent. Outside industry is a major point of contention.
Political	Commission of Communal Goods	Commission of Communal Goods Council of Elders Society for Social Solidarity
Conservation	Little participation in research projects and few students at the university. Active in ecotourism and other conservation-oriented projects.	Students at the nearby university are pivotal to research projects. Critical of programs and regulations of forest resource management.

In general, Ayotitlán and Zacualpan present contrasting community dynamics. Ayotitlán is active in promoting their indigenous identity, although they are an ejido and not an indigenous community. Through community development programs that emphasize environmental stewardship, they are demonstrating their trust and acceptance of non-community members. We have found that the more dynamic decision-making process was manifest in the community of Ayotitlán where they have the lowest socioeconomic status in the Reserve, and are considered to be a more "traditional" community. Households in Ayotitlán rely on fewer modern amenities such as televisions and refrigerators, which are common in Zacualpan households. By contrast with Ayotitlán, Zacualpan, the more "modernized" community in this study, demonstrated a more direct and congruent decision-making process.

Though we have not focused on the role of outside industries in this paper, there are fewer positive relationships with major industries in Ayotitlán than in

Zacualpan. Most importantly, perhaps, in the ways that outsiders perceive these two communities is that Ayotitlán is characterized as a welfare-dependent community, and therefore its community members are called "lazy." This is not the case in Zacualpan, however, where an increasing majority of households have family members working in low-level paid positions in neighboring cities. We contend that it is the perception of Ayotitlán as lazy and poor that contributes to outsiders' view that the community is generally more difficult to work in than Zacualpan is.

The political processes in Ayotitlán are also layered and more complex than the system of a single elected leadership body in Zacualpan. To some outsiders the Ayotitlán leadership structure is interpreted as disorderly and disjointed. Yet the multi-layered political structure in Ayotitlán may reflect very important dynamic and democratic processes of representation in the community. Some might contend that it would be simpler, in fact, for individuals in Ayotitlán to be passive recipients of "modernity" via state-sponsored programs and initiatives. In Zacualpan, the decision-making process is linear and simple by comparison. These same outside agents have attributed the ease of interaction with members of outside stakeholder groups to the relatively higher level of "modernization" in Zacualpan.

Individuals in both Ayotitlán and Zacualpan frequently have differences of opinion, as is true of people in general. In Ayotitlán they have clear mechanisms for discussing differences of opinion (most importantly at meetings of the Council of Elders). Also, there are community celebrations that act to periodically diffuse tensions in Ayotitlán. Distinctively, the community of Ayotitlán challenges relatively powerful outside institutions. They are not passive when dealing with outside stakeholders and they do not seek ways to "streamline" nor simplify their internal processes of decision-making for the convenience of outsider stakeholder groups. They do this to assert their rights as community members and human beings—they demand that ethical research and realistic policies for forest resource management be employed. By not allowing outsiders to enter easily they are protecting their community. This does not mean that the outsiders are trying to do them harm, nor that the community feels that. Rather, it suggests that they are not complacently accepting of whatever outside stakeholder groups suggest or promote.

The attitudes toward conservation are also important in both of the study communities because of the underlying mandate of the Reserve. While there are fewer active conservation-oriented research projects in Zacualpan, there are other manifestations of the community members' goals to conserve. Specifically, the riverside park demonstrates a progressive attitude toward their natural resources wherein they seek to simultaneously maximize their goals of making a profit, protecting the natural environment, and educating others in the region about the local natural and cultural resources. In Ayotitlán, however, we have not seen any successful proposals to promote ecotourism. Rather, there continues to be a proliferation of natural resource research in Ayotitlán. As community members in Ayotitlán actively work the land and their *milpas* (a tradition of growing maize in a polyculture and crop rotating system which has persisted for more than three thousand years in Mexico), regulations on their agroforestry activities have been a point of contention and disagreement with some outside stakeholder groups (namely the Management of the Reserve).

The perception of Zacualpan as being conservation minded and Ayotitlán being adverse to conservation would not be a fair conclusion to make. However, it is often the case that Ayotitlán is characterized as less interested in conservation. This perception is proliferated by the misunderstandings and misconceptions held toward Ayotitlán by many members of outside stakeholder groups.

Many questions remain about the process of sustainable development as programs are designed and carried out. Why are some communities better able to overcome differences of opinion and arrive at consensus? Is the consensus they arrive at truly reflective of shared goals, or is it simply a byproduct of disenfranchised and powerless community members? Studying and understanding local experiences and perceptions of sustainable development and conservation can help to answer these questions (Gerritsen 2010). As opposed to simply considering the outcomes, it is essential to view the dynamics that lead to such outcomes, since conservation and sustainable development are ongoing processes—persistent struggles (Long and Long 1992). Conservation and sustainable development are commonly measured with quantitative inventories of biodiversity and living standards, respectively. Yet mere conservation of biodiversity may not lead to success in instances where the process of conserving deteriorates cultural heritage (Brechin et al. 2002).

Social scientists and others in the area of natural resource management need to produce more critical analyses of the processes relating to project implementation in order to truly assess what has worked and what has not (Wilshusen et al. 2002). The local historical, social, political, and biological contexts are always very important components of how processes of community-based participation in resource management unfold.

Brechin and colleagues (2002) proposed six key elements of the social and political processes of community-based conservation programs: human dignity, legitimacy, governance, accountability, adaptation and learning, and non-local forces. Each of these aspects of the process of conservation and sustainable development initiatives has an impact on the outcomes of the projects.

Sustainable development must be community-based, participatory and empowering to the local actors. Though some have argued that community-based models of conservation are impractical in certain instances, Brechin and colleagues have stressed that conservation is essentially social action (2002). Ayotitlán has demonstrated the capacity to not be complacent, to express differences of opinion with outsiders, and to actively participate in determining the standards of local research and forest management. Under the rubric of "human dignity" Brechin and colleagues comment that involvement of community members throughout the process is an essential component that can help in seeking social justice (Brechin et al. 2002).

To properly evaluate the success of sustainable development, we need to look beyond abstractions, such as socioeconomic status, and also take into consideration the social and political goals of community members. When looking at the community members' self-portrayal and manifest objectives in Ayotitlán, it can be seen that they are successfully demonstrating what for them is most important: they are an indigenous community where they engage in the ongoing struggle to advocate for their own sustainable development needs and goals. There has been some contention in conservation literature as to whether or not local actors are appropriate deciders of resource management, however case studies have repeatedly asserted the importance of local decision-making and empowerment (Brechin et al. 2002; Ostrom 1990).

The legitimacy in the process of sustainable development programming is strengthened by having accountability throughout it (Brechin et al. 2002). Accountability is critical to the ongoing feedback cycle in program development in terms of the responsibilities, rights, duties, and assessments of outcomes (Brechin et al. 2002). The insistence in Ayotitlán of having copies of materials from research conducted in their ejido is important literally and symbolically. First, it gives them access to the information being collected in their community. Second, it contests the notion that they are somehow unable to appreciate scholarly and scientific work. Indeed, within academia there is a struggle to make research more accessible to the public, in general, and to the communities it is based on.

Furthermore, the involvement of students from Ayotitlán in research projects is also a complex form of benefit to the community. It puts the students in an advantageous position in terms of being able to collect data for their own theses and projects. It facilitates the higher education and success of indigenous people in scientific and academic careers. However, it also puts these students in a position where their ethnicity may potentially be used as a tool by lead investigators. They can be a convenient and economical source of "labor" for lead investigators, and they can make the process of obtaining consent a virtual non-issue. Indeed, this is the tradition in academia wherein those lower in the power hierarchy go through the process of training and inculcation into their respective fields. In cases where lead investigators presented their proposed research to the community, they also made a point to emphasize that the majority of research field workers would be students from the community. Thus, if the community prohibited the research project, those students would not have had the paid work and academic experience. Inevitably, and with good reason, the communities supported these students and accepted their roles as researchers.

Related to the aspects of legitimacy and accountability is the importance of governance and the structure of decision-making (Brechin et al. 2002). The level of democracy and promotion of individualism within a communal setting as seen in Ayotitlán should be respected. Far from being a complacent population, these community members are actively engaged in decision-making at a local level through various mechanisms, including the Ejidal Assembly, the Council of Elders, and involvement with social movements such as the EZLN. They are empowered and acting democratically to determine their own future.

We need to also be concerned with what the outcomes of sustainable development projects are, and if there has been appropriate feedback throughout the process of implementing the projects. Brechin and colleagues call this component of the conservation and development process "Adaptation and Learning" because there is no cookie-cutter model for conservation programming (2002). As the planning, implementation, and evaluation phases unfold there should be ongoing assessments and adaptations to the program. Immediate changes in some communities, specifically improvements in overall socioeconomic indicators, are not the underlying goal of sustainable development. Short-term benefits should be a central concern, but the long-term outcomes should have the highest priority. Furthermore, the engagement of community members in locally meaningful and beneficial activities that promote sustainable use of forest resources will have a longer-lasting impact. The support of local students in higher education is one such benefit. The long-term benefits of having a higher level of educational achievement among community members cannot be understated. Zacualpan has also experienced successes in terms of sustainable development. In our assertion that Ayotitlán is not entirely unsuccessful we do not imply that Zacualpan is. Rather, we used the example of Zacualpan to demonstrate the contrasting types of community dynamics and decision-making structure.

The impact of non-local forces, particularly policies and influences of large-scale industry, need to be taken into account in programming conservation and development projects (Brechin et al. 2002; Ostrom 1990; Wilshusen et al. 2002; Wilson and Thompson 1993). The underlying goal of ICDP is to promote positive social change through sustainable, community-owned initiatives. In order to truly achieve these goals, political action is necessary. The representatives of the Management of the Reserve are in a critical position to help guide political action, although political action has been a key activity of members of the IMECBIO, as well. Though it is also important to have these outside institutions advocating for policy changes, it is even more important for the communities themselves to be empowered to act on their own behalf in the political sphere. This very often requires the implementation of community-level organizations and access to resources.

Ayotitlán may be better situated to promote policy change since they have demonstrated a willingness and capacity to say "no" to outsiders. Furthermore, within the community the existence of various political perspectives is a manifestation of involvement and concern by community members. Indeed, we would find it very odd to see a community where everyone seemed to agree all of the time. As one of the commissioners of an ejido stated, "If there weren't problems, this wouldn't be an ejido." This sentiment reflects the inherency of conflict and tension in communities where goods (here, the forest) are owned collectively. This is why the decision-making process necessitates various levels and institutions, both within the community and outside.

Notes

1. This chapter is based on our article, "Community Development and Outsider Intervention in the Context of Protected Areas: Case Studies from the Sierra of Manantlán Biosphere Reserve, Western Mexico" which we published in 2011 in *Ecological and Environmental Anthropology* 6(1):10-25.

2. In this chapter, we also incorporate the research experiences and insights of Peter R. W. Gerritsen who has been conducting research in the Reserve since 1993. We use the plural voice throughout this chapter except when referring directly to other aspects of Olson's work presented in this book, for which she uses her singular voice.

3. Issues of tenure, privatization, agrarian reform in Mexico, and globalization are all implicated in these processes (Thompson and Wilson 1994).

Chapter 6
Indigenous People and Development Programs

This chapter addresses the production of value in community-based conservation projects within a Mexican biosphere reserve.[1] The discussion is situated in the discourse on community-based sustainable development strategies. In Mexico, the Biosphere Reserve is an important institution that reflects a national paradigm for conservation and development wherein cultural heritage is valued and prioritized (Halffter 1995). Some social scientists have contended that there are economic and political ideologies tied up in ecotourism interventions (Stronza 2007; Page and Dowling 2002b; Fletcher 2008; Page and Dowling 2002a). Scholarship on the role of ecotourism in sustainable development and international conservation has emphasized the role of neoliberal markets and capitalist ideologies in the growing area of ecotourism (Fletcher 2009; Argyrou 2005; West 2006; Alcorn 2001; Johnston 2006; Igoe and Brockington 2007). Furthermore, economic perspectives are essential to inquiries because ecotourism is an industry where market measures are a primary indicator of results.

The chapter explains the background of social and cultural studies of ecotourism in the context of conservation and sustainable development, and surveys relevant economic perspectives to arrive at a combined model of Weberian perspectives on rationality and Kapp's notion of social costs (McGehee and Meares 1998; Berger 2008). To exemplify the range of ecotourism projects along the continuum of the mixed model, two examples from the Reserve are presented illustrating alternate interpretations of motivations for ecotourism following the model. Each of the examples of ecotourism projects reinforces the need for situational rather than universal models of the creation and appreciation of value in community-based conservation (Wilshusen et al. 2002).

Sustainable Development and Conservation

Forest resource management and conservation is a process of social action and involves participants at local, national, and global levels. In all types of community-based conservation and development initiatives, the most important consideration is the "how" and not necessarily the "what" (Brechin et al. 2002; Chambers 1997). Social scientists and others in the area of natural resource conservation need to produce more critical analyses of these processes in order to truly assess what has worked and what has not. The local context—historical, social, political, and ecological—is always a very important component of how processes of community-based conservation unfold. In all situations, the political processes are layered and complex, reflecting the very important dynamic democratic processes of representation in communities. Brechin and colleagues proposed six key elements of the social and political processes of community-based conservation programs: human dignity, legitimacy, governance, accountability, adaptation and learning, and nonlocal forces (2002). Each of these aspects of the process is important in shaping the outcomes of conservation and sustainable development initiatives.

Sustainable development must be community-based, participatory, and empowering to the local actors in order to succeed (Pfeiffer and Nichter 2008). Though some have argued that community-based models of conservation are impractical in certain instances, Brechin and colleagues have stressed that conservation is essentially social action (2002). Under the rubric of "human dignity," Brechin and colleagues comment that involvement of community members throughout the process is an essential component that can help in seeking social justice (Brechin et al. 2002).

Legitimacy in the process of conservation and sustainable development is strengthened by accountability. In terms of responsibilities, rights, duties, and assessments of outcomes, accountability is critical to the ongoing feedback cycle in program development. Related to the aspects of legitimacy and accountability is the importance of governance and the structure of decision-making (Brechin et al. 2002).

Researchers also need to also be concerned with what the outcomes have been and if there has been appropriate feedback throughout the processes. Brechin and colleagues call this component of the conservation and development process "Adaptation and Learning" because there is no platonic form model for conservation programming (2002). As the planning, implementation, and evaluation phases unfold, ongoing assessments and adaptations to the program are necessary to ensure the procedural goals and outcome goals are being realized. Immediate changes in some communities, specifically improvements in overall socioeconomic indicators, are not the underlying goal of sustainable development. Yes, short-term benefits must be a central concern—but the long-term outcomes are of greatest concern. Furthermore, the engagement of community members in locally meaningful and beneficial activities will have a more lasting impact.

The impacts of non-local forces, particularly policies and influences of large-scale industry, need to be taken into account in programming conservation and development projects (Brechin et al. 2002). The underlying goal of integrated conservation and development programs is to promote positive social change through sustainable, community-owned initiatives. In order to truly achieve these goals, political action is necessary. Pfeiffer and Nichter (2008) call for anthropologists to continue as *cultural brokers* in these growing instances of sustainable development programming. Though it is also important to have outside researchers and institutions advocating for policy changes, it is even more important for the communities themselves to be empowered to act on their own behalf in the political sphere. This very often requires the implementation of community-level organizations and access to resources.

In Mexico, social movements have sought to enforce accountability in the processes of natural resource management and land reform. The Zapatista National Liberation Army (Ejercito Zapatista de Liberación Nacional, or EZLN) has a role in morally, politically, and socially representing the needs of the indigenous and peasant populations in Mexico. The EZLN gained international attention in Chiapas 1994 when its members made a dramatic demonstration, seized a town, and demanded that the Mexican national government fulfill its promises to the indigenous peoples—particularly in serious agrarian reform that would enable indigenous communities to have better livelihoods (Collier and Quaratiello 2005). The EZLN is one example of the form of social movement inspired by indigenous rights in contemporary Mexico.

The lower socioeconomic status that characterizes communities within the Reserve should not be seen as an indicator of failed sustainable development. It would be more beneficial to assess whether or not the social and political goals of community members are being met through engaging in the processes of sustainable development. There has been some contention in conservation literature as to whether or not local actors are appropriate deciders of resource management; however successful case studies have repeatedly asserted the importance of local decision-making and empowerment (Brechin et al. 2002). Local-level decision-making processes within communities are dynamic, give-and-take negotiations between community members and other stakeholders. Across communities within the Reserve there are different opinions and goals illustrated through local decision-making processes and negotiating with outsiders.

Integrated Conservation and Development Programs

Sustainable development should be seen as promoting conservation of the environment alongside improved socioeconomic conditions (Anand and Sen 2000). Because of this dialectical, grassroots emphasis, it has been lauded as a strategy for improving the quality of life in underdeveloped nations, yet it has also been criticized for repeated failures (Chambers 1997; Ferguson 1994; Israel et al. 1998). Sustainable development promotes the management of local resources in ways that directly benefit the local actors. However, designing sustainable development programs has proven troublesome for a variety of reasons, including that it has resulted in inequitable access to economic resources, lack of local ownership of programs, inappropriate policies of resource governance, and insufficient infrastructure to maintain programs over the long term (Minkler 1997; Wallerstein 2006; Gerritsen 2002).

As a response to these difficulties, Integrated Conservation and Development Projects (ICDPs, also discussed in chapter 3) have been promoted. ICDPs can encompass a variety of applied action research designs, including community-based natural resource management, adaptive management, or community-based conservation. ICDPs may not always entail research, but nonetheless are often designed to be participatory through the use of adaptive and iterative learning processes. ICDPs should reflect processes that aim to develop local capacities, empower participants, and (if research is entailed) balance research and action throughout the stages of the project (Israel et al. 1998; Wilshusen et al. 2002; Pfeiffer and Nichter 2008).

The process of ICDP implementation is just as important as the outcomes; it should involve a participatory procedure that emphasizes the centrality of relationships between individuals, groups, institutions, and their environments (Pfeiffer and Nichter 2008; Wilshusen et al. 2002). ICDP aim at empowering local actors by reinforcing the value of local culture and by providing avenues for the investment of outside funds that are managed locally (Pfeiffer and Nichter 2008). Rather than representing a unidirectional flow of resources, technology, and information into developing areas, there should be an ongoing exchange of ideas that is a dialogical learning process. Communities need to be enabled to take action on their own behalf in order for the development and improvements to be sustainable, otherwise interventions or policies may not be maintained over time (Wilshusen et al. 2002).

The essential goal of sustainable development in Mexico has been to promote both the immediate and the long-term benefits of natural and human resources for local people (Halffter 1995). In theory, the goals have been to empower indigenous and underprivileged communities to improve their socio-economic status through the managed exploitation of their human and material resources (Oldfield and Alcorn 1987). Because of this grassroots emphasis, sustainable development has been both lauded as a strategy for improving the quality of life in underdeveloped nations and criticized for repeated failures (Chambers 1997; Ferguson 1994; Israel et al. 1998). Sustainable development promotes the utilization of local resources in ways that should directly benefit the local actors in the immediate and long terms. However, successfully designing sustainable development programs has proven troublesome for a variety of factors, including: inequitable access to economic resources; lack of local ownership of programs; inappropriate resource governance policies; and insufficient infrastructure to maintain programs over the long term (Chambers 1997; Ferguson 1994; Pfeiffer and Nichter 2008).

Community-based conservation is a category of resource management strategies that overlaps with other participatory approaches to sustainable development. In participatory, community-based conservation and development projects, the goal is to have the community members be the lead actors in conservation and development programs—working side by side with researchers and policy makers (Ferguson 1994; Chambers 1997; Martinez et al. 2006; IMECBIO 2000; Pfeiffer and Nichter 2008). Co-management and participatory approaches are seen to be a compromise between using strictly local, or strictly top-down, approaches to natural resource management (Hoffman 2009). These types of community-based programs should be seen as holistic processes that emphasize the centrality of relationships between individuals, groups, institutions, and the local environments (Wilshusen et al. 2002). These types of community-based projects empower local actors, reinforce the value of local culture, and provide avenues for the investment of outside funds that are managed locally (Minkler 2004; Pfeiffer and Nichter 2008). Rather than being a unidirectional flow of resources, technology, and information into developing areas, community-based participatory research approaches are supposed to be ongoing processes of idea exchange and co-learning, but researchers have questioned the reality of this objective (Brechin et al. 2002; Ferguson 1994; Fletcher 2009; Haenn 2002). Sillitoe (2007) has articulated the hybridization of local, indigenous knowledge with globalized science. In ideal situations of community development programming, there is a cyclical and dialogical exchange of information between locals and outside agents. He asserts that sustainable development programs must include both local and scientific knowledge in the tool kit in order to most efficiently utilize human and natural resources (Sillitoe 2007).

Communities need to be enabled to take action on their own behalf in order for development and improvement programs to be sustainable and conservation to be meaningful (Björk 2000; Pfeiffer and Nichter 2008; Wilshusen et al. 2002). One particular problem facing many community-based conservation projects in developing areas is the need for economic opportunities. Anthropologists have noted the imperative for conservation goals to prioritize the need for economic development in marginalized communities (Alcorn 2001; Ferguson 1994; Gerritsen 1998; Igoe and Brockington 2007; Honey 2008). Sustainable development interventions hope to achieve these complementary goals through the integration of the social, economic, and environmental spheres (UNDP 2003).

Alternate forms of tourism have (re)emerged as part of a movement to promote heightened respect for peoples and environments. Ecotourism, geotourism, sustainable tourism, participatory tourism, nature- or community-based tourism, pro-poor tourism, green tourism, and heritage tourism are among some of the labels that have been used within this broad rubric (see Figure 6.1 for brief descriptions). The central ideals of ecotourism, which happen to overlap with these other types of tourism, are:

- Minimize impact,
- Build environmental and cultural awareness and respect,

- Provide positive experiences for both visitors and hosts,
- Provide financial benefits and empowerment for local people,
- Raise sensitivity to host countries' political, environment, and social climate. (TIES 1990)

The trend toward ecotourism approaches has been steadily increasing in Latin America as part of the sustainable development paradigm. They draw on many of the same core concepts, goals, and processes as other common ICDPs.

Category	Definition
Adventure tourism	A form of nature-based tourism that incorporates an element of risk, higher levels of physical exertion, and the need for specialized skill.
Ecotourism	Responsible travel to natural areas that conserves the environment and improves the welfare of local people.
Geotourism	Tourism that sustains or enhances the geographical character of a place its environment, heritage, aesthetics, culture, and the well-being of its residents.
Mass tourism	Large-scale tourism, typically associated with "sea, sand, sun" resorts and characteristics such as transnational ownership, minimal direct economic benefit to destination communities, seasonality, and package tours.
Nature-based tourism	Any form of tourism that relies primarily on the natural environment for its attractions or settings.
Pro-poor tourism	Tourism that results in increased net benefit for the poor.
Responsible tourism	Tourism that maximizes the benefits to local communities, minimizes negative social or environmental impacts, and helps local people conserve fragile cultures, habitats, and species.
Sustainable Tourism	Tourism that meets the needs of present tourist and host regions while protecting and enhancing opportunities for the future.

Figure 6.1. Typology of tourism.
Source: The TIES Global Ecotourism Fact Sheet, 09/2006, borrowed from the 2005 report by The Worldwatch Institute.

The ecotourism and other types of tourism emerging in the Reserve are interesting because of their complex integration of two dynamics: ecotourism and sustainable development, on a protected area. Explicit goals of conservation are woven throughout the discourse on sustainable development in the Reserve. Because this region is a protected area, it is subject to a more complicated decision-making structure, which is elaborated in chapter 5. There are limitations on the ways that individuals and families can exploit their surrounding environment, and novel opportunities are introduced in an effort to offset any reduced economic opportunities. Ecotourism is one of the common types of alternative economic activity found in the Reserve, but as my research showed, there are very different types of ecotourism. Some ecotourism may be perceived as a form of community empowerment, where community members actively promote expanding ecotourism, and in other instances ecotourism is done as a way of paying lip service to ICDPs.

The ecotourism and other types of tourism emerging in the Reserve are interesting because it is a protected area. Explicit goals of conservation are woven throughout the discourse on sustainable development in the Reserve. Because this region is a protected area, it is subject to a more complicated decision-making structure. There are limitations on the ways that individuals and families can exploit their surrounding environment, and novel opportunities are introduced in an effort to offset any reduced economic opportunities. Ecotourism is one of the common types of alternative economic activity found in the Reserve, but as my research showed, there are very different types of ecotourism. Some ecotourism may be perceived as a form of community empowerment, where community members actively promote expanding ecotourism, and in other instances ecotourism is done as a way of paying lip service to community members while pursuing other market and conservation goals.

Economic Models, Social Values, and the Discourse of Ecotourism

Max Weber's *formal rationality* applies to structured business decision-making in a market system where the goal is to maximize returns. Max Weber (1864-1920) was a German philosopher and political economist; he recognized that individuals (often acting in groups) also value qualities not accounted for in a strict market sense. *Substantive rationality*, then, was Weber's notion of the real societal conditions that frequently underlie formal rationality (Aron 1998). John Dewey (1859-1952) critiqued Weber for focusing these rationalities so heavily on the output goal (Berger 2008); Weber was not paying close attention to the social processes surrounding the creation of value and decision-making. Weber articulated substantive rationality as a theoretical tool, since his main goal was to promote a social science that was positivist and could advance testable hypotheses (Weber [1904] 1949).

Karl William Kapp (1910-1976) was a German-American economist and one of the founders of ecological economics. He saw that basic human needs are universally valued; hence, there was a universal basis for substantive rationality as proposed by Weber (Berger 2008). *Rational humanism* is Kapp's adaption of the biocultural theory of human needs and values; rational humanism expands the notion of substantive rationality by not limiting the concept to value (Berger 2008). Indeed, only some of the things human beings value can actually be quantified. Interpretive anthropology has seen this need for looking at human experiences through more inductive processes (Geertz 2000) as have other interpretive social scientists, including Weber himself (Berger and Luckmann 1966; Weber [1904] 1949).

Kapp's Theory of Social Costs may prove useful here because it provides a nuanced interpretation of Weber's substantive rationality and John Dewey's Pragmatic Instrumentalism (Berger 2008). Kapps' Theory of Social Costs is a useful tool in political ecology economics and has been applied to issues of sustainable development. According to Kapp, social costs are all of the direct and indirect losses to individuals or society that result from unrestrained economic activity (Berger 2008). Thus, the classic utilitarian approach—such as that of Weber's formal rationality—fails to adequately account for social costs. An important aspect of social costs is the notion that the "loss" could be avoided. Conservation literature suggests that we can at best hope to curtail biodiversity loss, but the human impact on biodiversity is not completely avoidable in settings where humans live (Alcorn 2001; Wilshusen et al. 2002). In addition, classic economic models see the structure of market relations linearly, and in the case of social costs to the environment, the losses are absorbed in a way that very often has no impact on traditional market dynamics (Fletcher 2009; Haenn 2002; Igoe and Brockington 2007). For example, the loss of an exotic bird species may indirectly impact some market interactions (such as bird watching) but would unlikely impact the cattle industry that led to the need to deforest. The supply and demand of beef cares little about the social costs of deforestation.

Environmental policies are inherently places where social costs and benefits are hard (if not impossible) to quantify discretely (Berger 2008). Additionally, the use of rational models where the focus is on monetary values is complicated in many of the world's most precious environments where local inhabitants live in poverty. Kapp saw that the willingness and ability to pay should not be applied to environmental decisions for these reasons (Berger 2008). However, Kapp was comfortable leaving social policy decisions to scientists and governmental agencies. Weber said that legitimacy in political leadership is accomplished by satisfying the social norms (Weber [1904] 1949); Dugan added that the legality of leadership is also fundamental to establishing legitimacy (Hoffman 2009). Care needs to be taken, therefore, throughout the processes of ecotourism programming to establish locally legitimate systems of leadership and governance. Governmental representatives and scientists may not fully appreciate local cultural values and can often be in conflict with indigenous rights movements (Haenn 2002; Higham 2007; West 2006).

The Weberian notion of rationality is clearly capitalistic. *Emic* notions of value (of culture, ethnicity, environment, forest resources, and so forth) are all essential to understanding the roles ecotourism may have in a sustainable development and conservation programs. *Emic* refers to the local system of knowing and the locally shared understanding of a concept; in this context, I am referring to the ethno-ecology of a community (Haenn 2002). The *emic* perspective is in contrast to the *etic* vantage point which is held by outsiders. In many ecotourism interventions, the dominant discourse comes from the West (the *etic* perspective [Fletcher 2009]). In light of existing critiques of *etic* perspectives, it is possible that the Weberian framework may not be fully able to grasp the *emic* aspects of value (Haenn 2002; Igoe and Brockington 2007).

A major shortcoming in many existing studies of ecotourism and sustainable development has been the reliance on outsider categories of meaning—a discourse of ecotourism that Fletcher (2009) has defined. Fletcher points out that ecotourism is an ideology in addition to a market strategy for economic opportunity—an ideology which impacts local populations but is initially foreign to them (Fletcher 2009). Significantly, there is a specific demographic of ecotourists who are the planners, operators, guides, and clients of ecotourism: white, middle-class, professional males (Fletcher 2008, 2009). In ecotourism, the priorities are self-expression and quality of life (Inglehart 1990).

Ecotourism is also a rural sustainable development strategy that needs to have local ownership (Agrawal and Gibson 1999; Honey 2003; Mowforth and Munt 2003), and be related to conservation internationally (Igoe and Brockington 2007). Honey (2008) refers to this notion that ecotourism needs to have local ownership as the "Stakeholder Theory" because the underlying assumption is that if there is value (mostly monetary) in the natural resource, then people will be motivated to conserve it to maximize their receipt of that value. The emphases here are on market integration and material value, which are both important components of neoliberal strategies (Igoe and Brockington 2007). In the places where ecotourism is promoted—especially as part of a conservation and sustainable development project—there is frequently an unintentional promotion of a neoliberal and capitalist ideology (Fletcher 2009; Haenn 2002; Honey 2008; Igoe and Brockington 2007).

Indeed, the mere promotion of ecotourism as a poverty alleviation strategy is neoliberal (Fletcher 2009). Furthermore, the notion that natural resource management can be market regulated is capitalistic (Alcorn 2001; Fletcher 2009; Wilshusen et al. 2002). In ecotourism, there is an emphasis on competition and privatization of resource ownership, which are core components of capitalism; the prominent model of ecotourism commodifies landscapes (Fletcher 2009). Thus, when ecotourism is promoted locally what is really being promoted is a neoliberal ideology (even if this is not an explicit goal).

Some scholars have pointed out that market approaches to ecotourism can often be counterproductive to conservation (West 2006; Higham and Carr 2003; McCauley 2006). In Peru, Stronza (2007) found mixed results. Some people saw the value in the environment as an incentive to continue promoting conservation techniques, but others took the money earned from the ecotourism activities and went on to invest in other technological advancements that were not aimed at conservation (Stronza 2007).

Fletcher (2009) aims to describe the socio-cultural context of ecotourism planning and intervention. He suggests that there is a clear discourse, even if it is only implicit, that this discourse is based on the beliefs, values, and assumptions which are promoted during ecotourism projects (Fletcher 2009). Furthermore, though much of this discourse is unwittingly transmitted, many ecotourism projects emphasize "environmental education" which is actively imparting this ecotourism discourse (Fletcher 2009). In order to train and educate local ecotourism guides (to achieve the local ownership and involvement in the ecotourism) there is extensive education which actively imparts the discourse of ecotourism. Albeit, the big question that remains is how local populations engage with the ecotourism discourse (Appadurai 1996).

Ecotourism in Latin America

Theoretical perspectives on tourism in Mexico and Latin America have appeared with different types of analyses (Garcia and Nechar 2006). Weberian, symbolic interactionism, formalism, phenomenology, and ethno-methodology have guided analyses of tourism at the micro-level (Garcia and Nechar 2006). Because of the localized nature of ecotourism interventions, it is appropriate to think about micro-level analyses (Fletcher 2009). Higham points out that one of the primary challenges in studying "ecotourism" is the lack of a consensus on a definition (2007). The struggle to define ecotourism is ongoing and restricts some types of analyses of projects that take on the label of "ecotourism" (Fennell 2008; Higham and Carr 2003). In addition, there are a cluster of subtypes of tourism that have overlapping intentions. Ecotourism, geotourism, sustainable tourism, participatory tourism, nature- or community-based tourism, pro-poor tourism, green tourism, and heritage tourism are among some of the labels that have been used within the broad category of these activities (TIES 2004). Ceballos-Lascuráin notes that "in reality ecotourism fits within the larger concept of sustainable tourism" (1998, p. 8). The central ideals of ecotourism, most of which overlap with these other types of tourism, include:

- Minimize impact
- Build environmental and cultural awareness and respect
- Provide positive experiences for both visitors and hosts
- Provide financial benefits and empowerment for local people
- Raise sensitivity to host countries' political, environment, and social climate. (TIES 2008)

Ecotourism is not easily operationalized as a social activity because we cannot define it, according to some scholars (Björk 2000). However, other scholars have asserted that through the process of defining the forms of ecotourism we can engage in meaningful analyses of the phenomenon (Fennell 2008; Higham and Carr 2003; Higham 2007).

Significantly, the concept of ecotourism originated in Mexico. In the 1960s the term emerged and originally had a focus on natural and archeological sites. Hetzer, a Mexican ecologist, introduced ecotourism and indicated that it should:

have a low impact on the environment; embody respect for indigenous cultures and not attempt to change them; provide economic benefits for local communities; and have recreational value for visitors (as cited in Higham 2007). Therefore, Hetzer saw that ecotourism could aid conservation, embody respect for local peoples, and maximize economic development—which are still common aspects of the term as used in contemporary parlance.

Coming into the 1980s, a number of global environmental issues shifted the focus of ecotourism to include ecological sites. It was Ceballos-Lascuráin who coined the classic and most widely cited definition in 1987 (Ceballos-Lascuráin 1987). In the 1990s, discourse and analysis of ecotourism began to consider some of the implications of the many definitions of ecotourism as a concept and activity. In addition to the noticeable lack of comparative ecotourism studies, there are substantial examples in the literature of poorly planned and implemented ecotourism interventions (Björk 2000). There was a strong critique (Wheeller 1991) which noted that ecotourism was not a benign activity; Wheeller noted that ecotourism is not always "responsible" in all regards. Other scholars also began to point out the difficulties of ecotourism (Butler 1990; Boo 1990; Bramwell and Lane 1993; Wheeller 1995). But, in so doing they also found many of the merits of ecotourism (Butler 1990; Orams 1995; Butler and Hinch 1996; Hall and Boyd 2003).

Ecotourism can be lauded, first and foremost, as a strong effort to protect the natural environment (Higham 2007). This includes very often working to conserve endangered species. Employment is created by ecotourism projects, which may support poverty alleviation efforts, and also support community empowerment (Björk 2000; Higham 2007; Stronza 2005). In terms of regional development and making economic transitions for developing regions, ecotourism can play a very positive role (Higham 2007). Ecotourism can be a positive model for communicating and environmental education (Orams 1995). Ecotourism tour guides are central actors in propagating the discourse (Fletcher 2009; Salazar 2012).

There is still discussion of a uniform model or definition of ecotourism (Björk 2000; Higham 2007). There is no consistent, empirical evaluation of the impacts of ecotourism interventions. The trend toward ecotourism approaches has been steadily increasing in Latin America as part of the sustainable development paradigm (Fletcher 2009; Haenn 2002; Halffter 1995), drawing on many of the same core concepts, goals, and processes as other community-based programs.

Björk (2000) finds that comprehensive definitions of ecotourism can be too restrictive because they attempt to accommodate such a breadth of activities. Because of the difficulties inherent in defining and categorizing ecotourism, Björk suggests the use of a continuum in thinking about how to relate different ecotourism interventions (2000).

Instances from Mexican Biosphere Reserves have shown that timing and the relationships between key stakeholders can directly impact the success of a sustainable development program (Hoffman 2009). Ecotourism-type activities are essential aspects of sustainable development in indigenous communities, particularly in protected areas (Alcorn 2001; Halffter 1994). First, ecotourism is usually an explicit assertion of the value of indigenous culture and tradition, creating impetus to retain cultural heritage (Alcorn 2001). Second, these programs are often extensions of the global political and economic complexes that shape and constrain socio-economic realities of marginalized communities (Hoffman 2009). Community members are encouraged to take up ecotourism projects in order to pursue new forms of economic opportunities, but such projects bring explicit ideological structures with them—such as free market capitalism (Haenn 2002).

The ecotourism and other types of tourism emerging in the Reserve are also interesting because it is a protected natural area. Explicit goals of conservation are woven throughout the discourse on sustainable development in the Reserve. Additionally, being a protected natural area, there is a more bureaucratic decision-making structure (Olson and Gerritsen 2011). There are limitations on the ways that individuals and families can exploit their surrounding environment, and novel opportunities are introduced in an effort to offset any reduced economic opportunities (Gerritsen 2002). Ecotourism is one of the common types of alternative economic activity found in the Reserve, but as the two examples presented here will show, the processes of ecotourism can lead to strikingly different models. One ecotourism example may be perceived as a form of community empowerment, where community members actively promote expanding ecotourism, and in the other instance ecotourism is done as a way of paying lip service to community members while pursuing other market and conservation goals. The two examples that follow will demonstrate how specific ecotourism projects can be discussed using a Weberian-inspired framework (adapted from Garcia and Nechar 2006, and McGehee and Meares 1998). The primary areas of consideration in the examples are: type of natural environment; type of project; explicit project goals; leadership and ownership of the project; process of the intervention; paradigm or dominating ideology that guides the project.

Ethnography of Ecotourism in the Reserve

These two examples are based on my participant-observation from February 2007 through August 2008. I was actively engaged in nearby communities conducting census and other data collection, which allowed me opportunities to regularly visit Cuzalapa and El Terrero. While in Cuzalapa and El Terrero, I recorded field notes on my participation as an "ecotourist," as well as unstructured interviews with key informants, including: representatives of governmental agencies directly involved with the initiation and management of these projects; community leaders; local project administrators; service providers and workers; and visitors to the sites. My sampling strategy was purposeful, actively seeking out stakeholders in the ecotourism projects in each community, using a snowball technique where informants directed the researcher to other key players (there were seven complete interviews, four in Cuzalapa and three in El Terrero, which are the basis of this analysis). The themes focused on in these interviews were: economic and cultural interests; goals and objectives of community members; outcomes achieved; struggles throughout the process; and the nature of interactions between stakeholders. In the process of analyzing the results of the interviews, additional details and clarifications were obtained directly from key informants at the Management of the Reserve.

In chapter 5 I explained that there are a number of key stakeholders who are "outsiders" to communities in the Reserve. Government agencies, research groups, and industry can all be characterized as stakeholders in the Reserve conservation and the sustainable development arena. In the process of sustainable development and conservation programming, local-level decision-making is critical to successful outcomes (Chambers 1997; Wallerstein and Duran 2006; Wallerstein 2006; Wilshusen et al. 2002).

The resulting information puts together two examples of how the goals of the communities directly engaged in ecotourism projects have been shaped and (re)created. To understand the formation and expression of these communal goals and identities, the examples are presented with a focus on the social and historical process of the projects. Taking a start from a Weberian notion of rationality, the focus is on the actions of individuals in social organizations and institutions, all of whom occupy different positions within the sustainable development rubric. The process from project proposal to implementation and maintenance is highly dynamic and varies in important ways between the two examples. These examples illuminate nuances of actions at the local level that reflect different emphases of formal and substantive rationality in group goals with ecotourism projects.

FIRST EXAMPLE: Color de la Tierra

Color de la Tierra (Color of the Earth) is an indigenous women's group dedicated to the production of *artesanias* (artisanal crafts), all-natural coffee, and other local products for sale in regional markets. Theirs is a classic model for sustainable development because of the emphasis on the value of adult women as essential providers for the household (Gutierrez and Gerritsen 2009). Their model fits under the category "ecotourism" because individuals come, from Mexico and abroad, to buy coffee and tour the shade-grown coffee fields (TIES 2004).

Color de la Tierra is dedicated to the conservation of biodiversity and traditional culture in the Reserve, and one of the primary ways that they achieve this is through educating customers about: their all-natural products; the community of Cuzalapa and their own families; their work; and the local environment. Tour groups come (mostly from the coastal resort towns) to tour their coffee plantations and buy their coffee and other products locally hand-crafted. Other types of products that are produced by the women include (depending on seasonality and interest of women in producing these specific items): embroidered clothing, napkins, towels and other cotton accessories, honey, sauces, tostadas, jams, preserves, and pomades. When they sell their coffee, they are selling an ideology and perspective on sustainable development. Coffee, they point out, is a medicinal plant as well as a beverage. The holistic philosophy of this women's cooperative asserts that the environments where we live and the foods we eat are the basis of our health (Gutierrez and Gerritsen 2009).

Shade-grown coffee is much more sustainable form of cultivation than coffee grown in the sun and plays an important role in conserving biodiversity because it provides a habitat for various bird and mammal species as well as preventing soil loss. Examples of the types of trees that are used to produce this shade in Cuzalapa include: mango, avocado, lime, orange, and oaks, among many others. These fruit trees both help promote a healthy coffee crop and provide the women with other fruits from which they craft jams and preserves. There are no agrochemicals used, making their products all natural and chemical-free. In Cuzalapa, the tradition of growing Arabica coffee goes back to the colonial period; over half of their coffee plantations are more than one-hundred years old (Gutierrez and Gerritsen 2009).

The price of coffee collapsed in the 1990s, and following that crash many of the small parcels of shade-grown coffee were converted into more lucrative activities: being cleared for crop cultivation (maize) or livestock pasture. In 2001, a group of women from the community organized themselves as a cooperative to process and sell the coffee. They reasoned that no one was taking advantage of these remaining coffee fields, and rather than letting them go to waste or clearing them for cattle grazing, they could do their best to take advantage of that coffee and sell it at the best price they could in the region. They say that their purpose is to "improve the economic, social and environmental situations and to preserve the biodiversity of the coffee plantations."[2]

The women's group Color de La Tierra has been designed from the onset to build toward complete autonomy and independence; they do not want to rely on grant money from the university or government to be a successful business. To initiate the process, field representatives from the Management of the Reserve helped the women's group establish official membership standards and constitute themselves legally with the national government as an indigenous cooperative operating for profit in 2001 and 2002. This gives them a high degree of autonomy in terms of how the group and the business function. Other support and assistance for the group Color de la Tierra came from some of the researchers at the local university who helped solicit grants and start-up funds for equipment.

In 2008, Color de la Tierra was reportedly almost entirely independent and is faring rather well economically. Membership in the group varies seasonally, but they are averaging an active membership of more than twenty adult women from the community. They have completed a new workshop facility, the amount of products they produce is expanding, and they are receiving more frequent and larger groups of visitors. Data regarding the number of visitors was not collected for this case study, but one of the group leaders estimates that they receive more than two-hundred visitors from outside of the Reserve each year. In fact, they cannot keep up with the demand for their artisanal coffee, which continues to sell out. The women's group in Cuzalapa is an example of the type of sustainable development program that is emphasized in the Reserve by government and local university groups aiming to integrate research into community development programs.

SECOND EXAMPLE: Cabañas in El Terrero

The cabins and rustic tenting areas in El Terrero were created by the governmental office of the Management of the Reserve (but are maintained and serviced by workers from the community). There are hiking paths and guided nature hikes which are offered by a few of the local community members. The ecotourism project in El Terrero reflects the emphasis on conservation in the forest—an ideology preached by researchers from the local university and the Management of the Reserve—and fits under the category "ecotourism" because individuals from within and outside Mexico come to appreciate the natural environment and use the rustic camping and hiking in El Terrero.

In El Terrero, as of the end of 2008, there are various rustic campsites for tenting, three cabins which each accommodate four adult guests, and a lodge with furnished kitchen accommodating up to twenty-four adults. The cabins themselves are of an environmentally conscious design for efficient and conscientious use of sustainably harvested local timber. Guided hikes aim to educate about the natural environment and biodiversity. The area boasts a wealth of bird species, including some rare and exotic species, attracting all types of bird watchers—from amateurs to professionals. However, these are more than bird-watching hikes. Local tour guides educate about the history of logging and cultural exploitation in the community. They point out reforestation projects in various stages, and other ongoing economic activities. In this community they also have a locally owned and operated timber mill for the sustainable extraction of timber. The field representative who has collaborated with El Terrero reports that there are fewer than ten adults in the community who participate in the ecotourism activities, and they only receive around fifty visitors each year who are usually from other parts of Mexico (though some visitors are international).

The national program under which the cabins were constructed was sponsored by PRODERS (*Programa de Desarrollo Regional Sustentable,* or Regional Sustainable Development Program, which is now PROCODES, *Programa de Conservación para el Desarrollo Sustentable,* or Conservation Program for Sustainable Development) under CONANP (*Comisión Nacional de Áreas Protegidas*, or National Commission of Protected Areas) and provided the start-up funds to create these camping sites and cabins. They were constructed in 2003 and 2004 in an effort to promote ecotourism as part of a national sustainable development program. Upon completion of the project, a local ecotourism committee was formed to manage the ecotourism activities, but all reservations and logistical management are still handled by the Management of the Reserve, whose field representatives refer to themselves as "liaisons" in the process.

The positioning of the Management of the Reserve as liaisons, or intermediaries, in this community-based ecotourism project is problematic. They assert that they serve two primary functions: (1) screening and control of visitors to the area, and (2) making planning in advance a requirement for visitors and community members, alike. The goal is to ensure that the type of tourism that is being attracted is not destructive but rather harmonious with the goals of conservation. Yet it is not clear how that can be achieved through a reservation system. The potential visitors are advised of the ecological fragility of El Terrero, but once they arrive and are in the forests there is no real way of knowing that they are not littering, wandering off marked trails, or otherwise being destructive to habitats. Instead, this instance is an example of policy-based enforcement of conservation through strict protection, which has had very mixed results cross-culturally (Wilshusen et al. 2002). One thing is clear: many community members elect to not participate in this (and other) sustainable development activities managed by the Management of the Reserve. Only a handful of community members in El Terrero elect to participate in the ecotourism committee (membership varies by season but typically ranges between eight to twelve individual adults).

There is a visitor center located at the highway turnoff where visitors are asked to register and pay for the camping or cabin rental, as well as purchase locally crafted souvenirs. Again, the external governmental management group manages this visitor center. Thus, the community members provide the services to visitors, but are not as directly involved with the business aspects of the ecotourism activities as are the women in Cuzalapa.

Applying the Analytical Framework

Using a Weberian framework to analyze the two examples presented here helps to illuminate different points on the continuum from what is formally rational (adhering to an economic benefits model) to what is substantively rational (in which both ideological and economic goals are central points). The Weberian notion of *ideal rationality* is useful in discussing ecotourism's role in a sustainable development paradigm precisely because it causes us to use an economic model in which the ultimate goals are guided by our moral appreciation for culture, ethnicity, and the environment (Garcia and Nechar 2006; McGehee and Meares 1998). In particular, what aspects of culture and environment are being promoted, and concurrently what economic strategies are being promoted and valued?

Max Weber's formal and substantive rationalizations are the basis for the framework in this evaluation of the two forms of ecotourism in the Reserve. The two examples presented demonstrate the extremely different ways that ecotourism projects can embody aspects of the Weberian framework for rationality (table 6.1). Following McGehee and Meares (1998) I focus on six main characteristics in each project: environment, project, goal, leadership, process, and paradigm.

Environment

There is a contrast in the types of environments where the Color de la Tierra and Cabañas are located. Color de la Tierra is in a more humid, semi-tropical forested area which is readily accessed by a paved road. The climate and accessibility stand in contrast to El Terrero, where the Cabañas are located—which is not accessible year round and often requires four-wheel drive to climb the rocky dirt road to the high altitude pine forest. For the different types of environments, different types of ecotourism projects will perhaps attract contrasting types of ecotourists (Fletcher 2008). However, in both of these environments birding is a primary focus of the local ecotourism groups and some of the same exotic bird species (such as the Caracara) are present in both environments. As explained earlier, some social scientists see the variation in ecotourism projects (such as the vast differences in environments and climates) as a possible reason for not using "ecotourism" as a universal term (Björk 1997; Higham 2007).

Project

The ecotourism examples I described provide different types of recreation. A central challenge in ecotourism studies is defining what the label implies and delineating what types of recreation it includes. Nonetheless, in Mexico the term ecotourism has been used to describe both of these projects and has been self-applied by each of the projects. In Cuzalapa, Color de la Tierra give tours of their shade-grown coffee fields and emphasize the local cultural and environmental impacts of their women's cooperative. The income they derive comes from both the sale of coffee and handcrafts and the fees for the guided tours. In El Terrero, with the Cabañas, the main source of revenue is the cabin rentals and guided tours. There is also some sale of handcraft items, but these are seasonal. Thus, in each instance the tours are a focal source of income and a shared characteristic. As has been demonstrated, tour guides are explicitly conveying the discourse of ecotourism as they educate participants (Fletcher 2009). In both Cuzalapa and El Terrero, ecotourists are literally sold a message about conservation; sustainable development and community that embodies neo-liberalism (see Haenn 2002 and Fletcher 2009, for further discussion).

Goal

There is an explicit goal that is stated by the Color de la Tierra on their promotional materials and website: it is to improve the socioeconomic condition of the women and their families, as well as to promote environmental education and to promote sustainable land use practices. Once again, it is clear that a goal of this women's cooperative is to promote what Haenn (2002) and Fletcher (2009) have described as the neo-liberal discourse of ecotourism. The Cabañas in El Terrero are less explicit about their goals to promote the discourse, as their focus is on measurable environmental conservation and to provide some revenue to the community. Formal rationales apply to both the examples described because of an emphasis on revenue, although the Color de la Tierra are also working on ideological goals and thus have a large degree of substantive rationality.

Leadership

In Cuzalapa, the local women's group and other community members actively pursue rural tourism activities (Gutierrez and Gerritsen 2009), whereas the Management of the Reserve has been the driving force behind the ecotourism activities in El Terrero. The women of Cuzalapa are acting specifically to conserve their environment and cultural heritage, while striving to improve the economic position of their families. The individuals in El Terrero can be seen as promoting environmental education and awareness while they engage in sustainable forest resource extraction thorough their lumber mill. Both work in collaboration with the Management of the Reserve to achieve their sustainable development goals, but to different degrees. Whereas the leadership in Cuzalapa is completely local (and female), the leadership in El Terrero is predominately external to the community.

Process

There is also a stark contrast in the processes of the ecotourism projects in Cuzalapa and El Terrero. Whereas the women's cooperative was an organic, grassroots process, the Cabañas in El Terrero has had a top-down process. In Cuzalapa, the women initiated and directed the growth of the project, while the Management of the Reserve has been the designer and driver of the Cabañas in El Terrero. Interestingly, Color de la Tierra has reported very rapid growth and great success in their revenues, whereas the Cabañas is a slow-paced project. Here, it is interesting to see that the governmental institution is pushing for substantive rationality in El Terrero. It leads us to these classic questions of whom conservation serves and, more importantly, who is responsible for decision-making in conservation agendas? It is problematic when outside agents—in this case, the Management of the Reserve—are making decisions on how to conserve, especially since community members are responsible for dealing with the repercussions (and expected to receive benefits). As environmental concerns become increasingly salient in other areas of Mexico, we should be conscious of the meanings of, and motivations for, conservation by local peoples (Appadurai 1996; Fletcher 2009). Individuals and families actively seek ways to improve their socio-economic opportunities, and being on a biosphere reserve alters the types of options that they have.

Paradigm

The ideological goals—the varying degrees of both formal and substantive rationales evident in each example—are quite different. In Cuzalapa, the model of the Color de la Tierra is similar to a traditional model of sustainable development and emphasizes the a priori role of women as primary providers for the household (Sachs 1992). In El Terrero, the Cabañas project is very aligned with a common conservation model which focuses on restrictive zoning and protectionism, including the prominent role of governmental management (Wilshusen et al. 2002). Through these very different paradigms we can see a similar emphasis on substantive rationalities. The importance of the economic benefit is actually more salient in Color de la Tierra because one of the explicit goals of the ecotourism project is to promote a sustainable livelihood for women and their families. The substantive rationality is imparted by the Management of the Reserve who are working to conserve and preserve in El Terrero; in Cuzalapa the substantive rationality and ideological goals are espoused by the women themselves.

It is possible to interpret these two examples as optimistic, ends-justifies-the-means, examples of the social and economic impacts of ecotourism; the examples presented here may be examples of successful sustainable development and conservation projects in the Reserve. Simultaneously, the continuous struggle to expand the variety, quality, and quantity of handcrafted items sold in Cuzalapa, and the increasing number of visitors to El Terrero, are manifestations of economic and cultural globalization—the neo-liberal pressure (Haenn 2002). The demand for "natural" and ethnic "indigenous" handcrafted items increases disproportionately to the prices these communities are able to charge. There is an appropriation of what their cultural identity is supposed to be by outsiders, and an arguably non-equitable price that is often paid for those goods and services. The communities do not have the autonomy to utilize forest resources and offer their services in ways that could maximize their economic profits. It is presumed that attempts to do so would automatically lead to environmental degradation (see Fletcher 2009 for a discussion of the *colonial mentality*). The Management of the Reserve directly and indirectly restricts them to the expansion of ecotourism and the production of handcrafted items. Thus, the romantic-

Table 6.1. Summary of characteristics of two ecotourism projects: Color de la Tierra and Cabañas El Terrero.

	Color de la Tierra, Cuzalapa	**Cabañas, El Terrero**
Environment	Low cloud forest. Semi-tropical. Low-intensity agriculture and livestock tenure. Easily accessed by paved and maintained dirt road year round.	High altitude (2000m), pine forests. Timber extraction and some livestock tenure. Dirt road for access not regularly maintained. Need all terrain vehicles during rainy season for access.
Project	Shade-grown coffee tours, casual birding, sale of handcrafts and food products.	Cabins, guided tours including bird-watching hikes. Horseback riding. Some sale of local handcrafts.
Goal	To improve the socioeconomic condition of the women and their families. Promote environmental education and sustainable land/forest use practices.	Ensure conservation of the environment/forest. Provide supplemental income to community.
Leadership	Women's cooperative.	Management of the Reserve Community Advisory Board.
Process	Initiated by the women of Cuzalapa. Rapid growth. Sell out of coffee each of past four seasons. Planning to construct cabins and camping areas for overnight and extended visits.	Initiated by the Management of the Reserve. Slow growth. Construction of one new cabin/year on average. Not eager to develop new trails.
Paradigm	Classic sustainable development strategy—focuses on women as primary providers for household economy.	Classic conservation model—zone restriction and protectionism. Heavy influence of government management team.

cized, exotic "native" becomes a commercial good in a market that restricts the value of what would otherwise be focal commodities (the forest).

There are two primary perspectives on ecotourism projects such as those in question here. One, they can be viewed as a positive way of strengthening local pride, culture and environmental quality while bringing money into a marginalized community through the provision of tourism services (Alcorn 2001; Björk 2000; Higham 2007; McGehee and Meares 1998). Or, they can be seen as another form of exploitation of marginalized communities that reifies dependency on outside economies and promotes cultural imperialism through the globalization and marketing of indigenous identities (Fletcher 2009; Haenn 2002; McGehee and Meares 1998). The Mexican model of the biosphere reserve dictates that living cultures are part of the conservation and development effort (IMECBIO 2000; Halffter 1995); indigenous identities are quintessential to the goals of the Reserve.

Conclusions

This chapter proposes that concentrating on local, emic notions of value can lead to ecotourism strategies that simultaneously fulfill formally and substantively rational goals. Conservation projects stimulate the creation of value as much as they fortify existing cultural values. Situational models of conservation value production as opposed to universal models are better.

As Björk and others have often pointed out, "Lofty definitions describe an ideal, not a reality" (2000, p. 41) and this continues to be a challenge for those scholars and communities involved with ecotourism. Although it is not necessary for theoretical analyses and operational case studies to be using the same set of definitions, it would be helpful (Björk 2000). Focusing on the processes along which social, environmental, and economic values are created and reproduced allows observers to think about self-described ecotourism projects. The benefit here is also in the looseness of the model which allows us to think in terms of a continuum (Björk 2000). In ecotourism studies, there is neither a consensus on how to define nor evaluate the projects. Incorporating an economic rational in the model is imperative because ecotourism is a market strategy, however as this and other studies have demonstrated, there cannot be a strictly economic perspective (Fletcher 2009; Haenn 2002; Honey 2008; Igoe and Brockington 2007). As we continue to delineate and study the productions of value in ecotourism, thinking in terms of situational and not universal models will prove beneficial.

Notes

1. This chapter is based on my article, "Notions of Rationality and Value Production in Ecotourism: Case Studies from a Mexican Biosphere Reserve," which was published in 2012 in the *Journal of Sustainable Tourism* 20(12):215-233.

2. Their website, www.cafecuzalapa.com, provides background information in both Spanish and English.

Chapter 7
Ethnomedical Systems in Mexico

The daily realities of indigenous people in the Reserve are manifestations of religion, health, ethnicity, and social class which come together in the formation and recreation of identity in the Sierra of Manantlán. Mexican scholars like Nestor Canclini argue that the formation of identity in calculated decision-making is a manifestation of these ongoing processes—particularly in the area of health care (Canclini 1995).

This chapter explores medical pluralism in Mexico. Looking at the structure of the healthcare sector in Mexico and the Reserve leads to an overview of the ethnomedical systems prevalent in contemporary Mexico. I begin with salient descriptions of the general health status in Mexico, but recognizing that marginalized communities like the ones studied in this book are at the lower end of the national trends. There are many interconnections throughout indigenous communities in Mexico, they share many common goals and attributes. But, more importantly perhaps, the national government is beginning to pay more attention to its rural constituency with the realization that they are the guardians of traditional ecological knowledge, the backbone of Mexico as producers, and increasingly as consumers. The health needs of community members in the Reserve, and the services available to them, are related to the usage of medicinal plants.

Mexican Healthcare System

The Instituto Mexicano del Seguro Social (Mexican Social Security Institute, or *IMSS*) insures 90 percent of all healthcare services to the Mexican population. The mission of IMSS is to promote health and welfare to the working population and their families through the provision of health and welfare services—including economic and social needs (IMSS 2005). Achieving such a broad goal means structuring the healthcare system so that it will work in all of Mexico's rural, urban, impoverished, and developing areas. The IMSS is the administrative branch of the Ministry of Health that provides healthcare services and subsistence assistance to the population (González-Rossetti and Mogollon

2000; Ward 1986; Lassey et al. 1997). As an administrative body, the IMSS strives toward social justice and equitable access to healthcare and other basic needs.

The healthcare made available to the poor through the Ministry of Health is also noticeably inadequate (González-Rossetti and Mogollon 2000). The Ministry of Health oversees the provision of healthcare services and has a critical role in public health initiatives. The IMSS is an independent, parallel organization for delivery of healthcare to the taxpaying sector. The Ministry of Health is responsible for providing public health and primary healthcare services to the remainder of the population—whether or not the individual is covered by any other health provider. Ninety percent of all healthcare services in Mexico are rendered possible by IMSS (PAHO 2003). In the Reserve, the IMSS is not the primary source of healthcare services, except for those few individuals in Zacualpan who work in neighboring Colima. In the Reserve, poor farmers have very limited healthcare services.

IMSS acts as an administrative branch of the government making the process for implementing policies and programs politically charged. Researchers have pointed out that in spite of IMSS goals of equitably redistributing resources and access to these resources, the ability of certain individuals and groups to obtain better healthcare services depends largely on income, occupation, social class, and political bargaining power (IMSS 2005). Laborers and workers are particularly strong supporters of IMSS, since they are able to organize themselves in unions that enable them to make the system work to meet their needs (González-Rossetti and Mogollon 2000; Lassey et al. 1997; Ward 1986; Harrison 2001). Individuals and families whose primary source of income is through the informal sector, who likewise tend to be located in rural areas like the Reserve, have little or no access to healthcare—or can access healthcare services that are at best "second-rate" (González-Rossetti and Mogollon 2000; Lassey et al. 1997; Ward 1986).

The delivery of healthcare services requires a diversity of health professionals and health service centers. Throughout Mexico, individuals directly and indirectly involved with providing health services, as well as non-biomedical workers and administrators, comprise the largest union in the nation, the IMSS union, numbering more than 350,000 individuals (González-Rossetti and Mogollon 2000; Lassey et al. 1997; Ward 1986). The IMSS union has strong political bargaining power due to its integral role in providing healthcare to the nation. Since the 1930s, when organized labor unions became officially supported by the State, unions have been an important body advocating on the behalf of laborers at a national level. Even though the IMSS union has played a stronger role in politics historically, the last decade has seen a more independent voice emerging from it (Lassey et al. 1997).

There are four domains through which primary healthcare is delivered in Mexico: (1) private practitioners, (2) Social Security administered clinics, (3) federal clinics, and (4) specialized institutes (Lassey et al. 1997). The latter three of these domains are administered, and regulated, in varying degrees by the

Ministry of Health (and more directly through the IMSS). The private sector serves less than 7 percent of the population (Lassey et al. 1997; PAHO 2003). In areas like the Reserve, federal clinics—community health centers—are the most accessible form of healthcare. These are typically staffed by one physician, a minimum of one nurse, and perhaps one or two other health professionals (National Center for Health Statistics 1998; WHO 1998; Lassey et al. 1997). Local healthcare workers are trained to help provide basic healthcare services, such as administering immunizations, giving first aid, or other basic tasks. This system of community-based health workers has been advocated by the World Health Organization as a mechanism for making basic health services available to rural areas (Lassey et al. 1997; WHO 2004). Preventive healthcare is also a key focus of community-based health workers.

Biomedical Practitioners in Mexico

In 1984, the Ministry of Health conducted a survey that found there were four times as many traditional healers as physicians in non-urban areas (Lassey et al. 1997). Of those, nearly half were *parteras* (midwives). Most of these midwives receive an apprenticeship training, indeed traditional midwives continue to constitute the majority of midwives in Mexico (Lassey et al. 1997). Certified nurse-midwives and other professional midwives are practicing in Mexico but in significantly less numbers.

Training of physicians has declined in Mexico in the last decade. There was a surplus of physicians in the late 1980s, leading to a drastic decline in medical schools (Lassey et al. 1997). A survey in 1986 showed a severe underemployment of physicians, particularly in urban areas, which may have contributed to the decline in medical school enrollments during that period (Lassey et al. 1997). Medical school training in Mexico is very similar to that in the United States and most of Europe: four years of coursework, followed by one year of internship at a hospital, and finally one year of social service in a community clinic (Lassey et al. 1997). Upon completion of the social service, physicians move on to complete a residency—but these are not widely available, forcing some graduates to leave Mexico to complete their residency elsewhere (PAHO 2004).

The number of physicians and nurses serving the Mexican population is an important factor in the accessibility of healthcare services, though these numbers do not reflect disparities in the distribution of these resources. In 2004, the ratio of physicians was less than 16 per 10,000 habitants; and there were less than 11 professional nurses per 10,000 habitants (PAHO 2004). The numbers of trained practitioners is not as critical as the disparities in accessibility to them; in Mexico, as in the United States and many other countries, there is a shortage of doctors in rural areas. Urban areas provide access to biomedical technologies many doctors rely on to diagnose and treat, in addition to the other social qualities and lifestyle options that are considered desirable by many doctors and nurses. It is important to understand the distribution and structure of Mexico's human resources when discussing the shortages in the rural Mexican healthcare system.

Traditional healers (for example, *curanderos*, *naturistas*, or *parteras*) are an important aspect of the provision of health services, particularly in less urban areas where biomedical services are more sparsely distributed (Lassey et al. 1997; Ward 1986). Sociopolitical factors are influential in the provision of health services. It can be argued that there are conflicting interests in the macro-level policies and programs of organizations representing the North in public health services, as well as in national biomedical programs. Simultaneously, there are local and individual goals of identity assertion and formation coupled with the desire to have the best options available. While being "indigenous" and retaining traditions from forbearers is highly regarded among some Mexicans, these same Mexicans are living in a global era where exposure to outside markets and goods has strong effects on identity and preferences. In the next chapter, the perspectives of medicinal plant experts, traditional healers, and biomedical practitioners are presented in a forum that explores the desires to continue using MPK, and some of the resistance that is encountered.

Health Profile in Mexico and the Reserve

Maternal and child health are key indicators of the health of a community (WHO 1998). In the Reserve, the most common uses of the biomedical outposts are related to: childbirth; injuries; child health; and chronic lifestyle illnesses (like type II diabetes and heart disease). Mothers and children are the most vulnerable to preventable illness and are the most directly impacted by socioeconomic conditions (WHO 1998). In the Reserve, the census reported Zacualpan with the highest rate of live child births (3.1) with Telcruz (2.68) and Ayotitlán (2.35) at lower levels (INEGI 2005). The average rate of live child births for the entire state of Jalisco was 2.53 and 2.49 in the state of Colima that same year, putting Zacualpan well above—and Telcruz slightly above—regional averages (INEGI 2005). In Ayotitlán, the lower rates of childbirth may reflect lower health and economic status, or they also may reflect family planning programs. According to this same census, the reverse trend is seen: only 4 percent of Ayotitlán's population does not have access to health care, compared to 23 percent in Telcruz and 29 percent in Zacualpan. However, this is most likely a result of the way that the question was asked and what was being measured. In Zacualpan, many of the adults work in surrounding industries, but only a small portion of those laborers from Zacualpan have formalized employment that would qualify them access healthcare through IMSS. In Ayotitlán, very few individuals work outside the community and most access healthcare services through the local health clinic which offers free or low-cost services (mostly aimed at maternal and child health). Most families in Ayotitlán qualify for this local clinic; in Telcruz there is not a regularly staffed clinic and most individuals in Zacualpan have to access healthcare services through their employment or pay out of pocket in nearby Colima. There is only one full-time doctor who alone staffs the clinic in Zacualpan (with the occasional nurse who works only on immunization or drug distribution days) and there are two or three full-time doctors and usually five nurses who staff the clinic in Ayotitlán.

Table 7.1. Childbirths and health care services in the Reserve.

	Zacualpan	Telcruz	Ayotitlán
Proportion of live child births	3.1	2.68	2.35
Population without rights to health care services	494 (29%)	269 (23%)	26 (4%)

Source: Instituto Nacional de Estadística y Geografía (INEGI) 2005. ITER2005. Sistemas Nacionales Estadístico y de Información Geográfica.

Maternal and child health are important indicators of the general health status of a country, and rates of other chronic illnesses are not as reliably recorded since they frequently are not reported to biomedical health professionals. Most salient measures in developing countries with sizable marginalized segments of the population are the morbidity and mortality rates of mothers and children. The reported infant mortality rate in 1999 for national Mexico was 15 per 1,000 live births, primarily attributed to congenital malformation, influenza, and pneumonia infections (PAHO 2003). This rate increased to 21 deaths per 1,000 live births in 2003 (PAHO 2004). Hospitalizations in Mexico are overwhelmingly related to maternal reproductive healthcare. In 1999, 67 percent of all hospitalizations were for pregnancy and childbirth services. The maternal mortality rate for live births was 51 per 100,000 in 1999 (PAHO 2003), which is actually not a very bad statistic for a developing country. However, as with the infant mortality rate, the maternal mortality has increased to 64 maternal deaths per 100,000 live births in 2002 (PAHO 2004). It is troubling to see that maternal and infant mortality is increasing so noticeably over the period of a few years, causing concern for the trajectory of the Mexican healthcare system (Lassey et al. 1997; Ward 1986). In spite of dismal health trends for reproductively active women and their newborns, 68 percent of these women are using some form of contraceptive technology, and 91 percent of all childbirths are being attended by skilled practitioners (PAHO 2004). There is a discrepancy between an apparent adherence to preventive reproductive strategies by many Mexican women, and the lack of improved health outcomes for these women and their newborns, signaling a problem in the Mexican health sector.

Medical Pluralism in the Reserve

Injuries and chronic diseases are the most commonly reported ailments among adults in the Reserve. In the study sample, and based on interviews with nurses at the health clinic, the most common health problems for adults are: colds/flus, respiratory infections, injuries, diabetes, and hypertension—in about that order. For serious injuries, such as broken bones or "excessive bleeding" medical attention is sought. Radiographs are routine, when they can be afforded—even when the injury is clearly muscular. There is a clear perception that use of technology is directly related to a higher quality of healthcare.

Most of my study participants corroborated that most people in the Reserve will either treat first at home using popular medicine, but at times it is also common to use folk and professional medical systems simultaneously. In one instance, a man had got his hand caught in some heavy machinery and was badly injured. The doctor at the local clinic put in stitches and wrapped the arm to prevent it from moving or getting infected. There were no broken bones, but the lesions were deep and there may have been a fracture in the forearm. After a few days, the wound became badly infected and he returned with his wife to the clinic. They removed the stitches and reported that the infection would subside using some topical antibiotic creams that were prescribed. They continued cleaning the wound and using the prescribed treatment, but his arm swelled to double its normal size. A third time they returned to the health clinic to demonstrate the growing severity of the wound. They were told to "just wait and keep with the treatment. It will get better." The wife was enraged and they left. Getting home, she gathered some special plants from her household garden and cooked them. When they were cooled, she unwrapped the gauze around her husband's arm and applied the plants. "I used the plants to cure my husband," she proudly explained to me. I sensed that, in the future, neither she nor her husband would be so quick to rely on the professional medical system.

Mexico contains a large diversity of indigenous groups and ethnomedical systems. A general categorization has been that of *folk, popular*, and *professional* ethnomedical systems. *Folk* ethnomedical system is synonymous with traditional medical system and refers to the ways of promoting health that developed over the centuries in local cultures, practiced by non-bureaucratic specialists (*curanderos*). *Popular* ethnomedical knowledge is the most commonly shared cultural knowledge—it is what I captured using freelisting of medicinal plants. *Popular* ethnomedicine is what everyone knows; there is no social structure associated with this system of knowledge and access to this knowledge is not restricted. *Professional* ethnomedical system is the bureaucratic structure of specialists who are integrated into a broader social structure (Kleinman 1980b). Kaja Finkler has studied these categories of ethnomedicine in Mexico, using both rural and urban settings to capture the diversity that is often greatest between different types of communities. She begins by questioning how folk and popular systems are different, leading her to choose both rural and urban research sites: Mexico City and an unnamed rural community (Finkler 1994a,

1994b). Important themes in Finkler's research include: source of livelihood, migration, family structure, gender roles and relationships, ethnicity, development, religion, and both traditional and biomedical systems of medicine (2001). In her research, Finkler has sought to identify the impacts of globalization on health behaviors, postulating that the rural/urban divide will resonate with differences in less-developed/developed, traditional/modern communities in Mexico.

Finkler has been very interested in the roles that gender plays in sickness experiences, and how these relationships are reflections of broader societal trends. Finkler looked at the patterns of social interaction in terms of gender and gender ideologies, many of which stem from a history of Catholic influences (Finkler 1994b). The cultural expectations of women are central to the etiology of many of her informants' chronic illness episodes. Finkler highlights the lived reality of poverty as an underlying factor. Beyond this, however, are the complex cosmologies that directly influence how these women feel about themselves—how they are as mothers, wives, daughters, sisters, workers-producers, and community members (Finkler 1994b). Domestic violence, both physical and emotional, is a common struggle for these women. The biomedical diagnoses that these women have received are based on the clusters of symptoms that they report. However, the manner in which these women report their experiences of pain are not necessarily congruent with a Western biomedical model of illness. Furthermore, the manner in which these women seek out different therapeutic options is illuminating because it points to what each of these women suspect the origin of her illness to be. In bringing together the women's stories, Finkler employs the concept of "life's lesions" to conceptualize the physical harm that results from the cultural and social factors that affect these women's lives (1994). The combined social influences of gender and economic status are frequently found to be of central importance in health globally, and specifically in Latin America (PAHO 2004; Byron 2003). Finkler's work is particularly crucial because she demonstrates the how gender, religion, social class, economic activity, and health play out in daily lives of contemporary Mexicans.

Even in urban parts of Mexico, gender and social class have been found to be essential in understanding health and ethnomedicine. Valentina Napolitano (2002) uses ethnography to demonstrate the changing social dynamics in urban, contemporary Guadalajara, Mexico as part of the ongoing migratory patterns in Mexico. What aids in Napolitano's ethnography is the time she spends situating the current residents in the colonial past and settlement history. Migration is a central axis—as modernization and globalization have unfolded in the community, migration has been an ongoing necessity. The actual geography plays a role in the reasons for migration as the landscape becomes increasingly urban and farming decreases (Napolitano 2002). Napolitano argues for the importance of individual agency in the formation of identity through faith, suffering, and healing in the medical pluralism of contemporary Mexico (Napolitano 2002). The role of the Catholic religion is, again, found to be a primary influence. However,

Napolitano also links local medical pluralism to failed national health reform programs in Mexico. These programs aimed to expand the accessibility of biomedicine, but have at the same time undermined many components, and providers, of traditional medicine (Napolitano 2002). This resonates most with Luz from Cuzalapa. Paradoxically, the many shortcomings of ambitious health reform have, at the same time, reinforced reliance on traditional therapies.

As the influence and impact of biomedicine become increasingly more available throughout urban, less-urban, and rural Mexico, the survival of traditional healing systems (e.g., midwifery, herbalists, and healers) becomes more impressive. Rural communities desperately want to be part of the formal sector and pluralistic model of ethnomedicine in Mexico. The inadequate access to biomedical care in less-urban areas of Mexico has contributed greatly to negative health outcomes in childbirth and children's health. Poor women and their families throughout Mexico have a desire for equitable access to appropriate healthcare (Lassey et al. 1997; Ward 1986). There are not sufficient provisions to ensure that these rural people also have access to the complete spectrum of biomedical healthcare (Lassey et al. 1997; Ward 1986). There remains a question of equity in access to biomedical services, particularly in rural communities like Ayotitlán, Telcruz, and Zacualpan.

When professional medicine is not easily accessible, or not seen as beneficial to the health, people will not resort to it. These sentiments are expressed in the next section in the words of medicinal plant experts. When professional biomedicine is powerful, it is preferred. However, it exists as simply one option among many.

Traditional Healing in the Reserve

The presence of traditional ethnomedicine in the three study communities is embodied in individuals who are experts in it. The prominent sources of healthcare are allopathic medical clinics. Kleinman, and others, have found that most of the remedy-seeking process begins in the home (Garro 1998a, 1998b; Good 1987; Kleinman 1978, 1980a). To the extent that MPK is the basis of many home remedies, it is highly prevalent in the three study communities. Yet this book is not about sickness experience nor treatment pathways. I am interested in exploring the connections between socioeconomic status and MPK because of the great deal of intracultural variability in these areas. The historical and political trajectories in Mexico have shaped the ways that health services and economic activities are manifested in the Reserve. At the intersections of these relationships are not only sustainable development programs, but the individuals who are experts in health and healing. Experts in medicinal plants and biomedicine will most dramatically demonstrate current trends or relationships.

I now turn to those individuals regarded as traditional healers or experts in medicinal plants. As experts in specialized domains of knowledge, their experiences with medicinal plants are likely to be exemplary of important trends; the perspectives of traditional healers and biomedical practitioners are simultaneously reflections, and influences, of community members' perspectives. Looking at their life stories, I focus on how they became skilled ethnomedical healers and their perspective on the current state of health and healing in their communities. In this chapter, it becomes clear that (1) there is a great deal of heterogeneity between communities, and (2) the social construct of the traditional healer in the Reserve is an extension of the same political-economic processes shaping the cultural landscapes in this region of Mexico.

Traditional Healers and Other Health Service Providers

I interviewed five traditional healers and experts of medicinal plants in the Reserve. Of these, one was male (sixty-five years of age) and the other four were women between the ages of thirty-four and sixty-three. An additional three interviews were completed with allopathic doctors and a nurse, who were all males between the ages of twenty-three and thirty-three. Of those interviewed, all were from the Reserve except for the two medical doctors carrying out their social service at clinics in Ayotitlán and Zacualpan.

Table 7.2. Semi-structured interviewees.

Informant	Community of Residence	Age	Occupation
Consuelo	Ayotitlán	65	*Curandero* (curer, healer)
Celestina	Zacualpan	63	Homemaker
Claudia	Zacualpan	50	Homemaker
Lourdes	Zacualpan	50	Peasant, field hand
Luz	Cuzalapa	34	Artisanal coffee producer
Efren	Ayotitlán	33	Nurse
Pepe	Zacualpan (originally Guanajuato)	23	Medical Doctor
Rodolfo	Ayotitlán (originally Queretaro)	23	Medical Doctor

Below I provide vignettes and short narratives based my interviews with the various practitioners, healers, and medicinal plant experts. The purpose is to highlight the experiences of those interviewed in terms of understanding the reasons for using medicinal plants and where that knowledge comes from. Examining their stories and perspectives sheds light on the bigger processes in the Reserve; the larger forces of modernization and sustainable development resonate through all aspects of daily life for community members.

Zacualpan

Lourdes uses medicinal plants to heal herself and her family. She is a fifty-year-old woman living and working in Zacualpan. She tells me her story from her bed where she lies ill with a terrible cough and fever; she has been bedridden for some time now (at least two months) due to an acute upper respiratory infection acquired through repeated exposure to chemicals in the workplace. She works as a field hand at a large indoor planting facility where the chemical fertilizers made her ill. On her nightstand there are packages of medicines and pills given to her by a biomedical doctor.

In spite of her illness, Lourdes talks enthusiastically about the benefits of medicinal plants and reports that she uses them because they are natural. Lourdes observes that medicinal plants are "good for you and good for healing. There are so many useful plants all around, and besides," she says, "modern medicine just takes the useful stuff of plants and uses chemicals to manufacture them." Thus, Lourdes sees plants as readily available, useful, good for health, and serving as the basis of biomedical pharmaceuticals which have been contaminated by chemicals.

The avoidance of exposure to chemicals is the primary benefit Lourdes cites she obtains from using medicinal plants. The irony is that she is currently very ill (bed-ridden) due to exposure to agrochemicals at the industrial greenhouse where she works. Regardless, she is firm in her promotion of medicinal plants. Lourdes explains how she can help cure her family members and friends—they exchange plants, remedies, and when someone is sick they help each other out by going and finding the plants. This shared, communal responsibility for using the plants to cure is important. Also, she notes that medicinal plants are part of an ancient technology that is potent and reliable (more trustworthy than biomedicine). Finally, she comments on the last benefit of using medicinal plants: they are not nearly as expensive as modern drugs.

Lourdes began learning as a young girl when her grandmother (and mother to some extent) taught her about medicinal plants. However, she credits two training seminars she has attended for the bulk of her knowledge and expertise. These courses took place in other parts of Mexico (one in Colima and the other in Michoacán) and provided a practicum and handbooks on using natural methods of curing. She shares this information with other community members in Zacualpan. Yet Lourdes tells me that other people do not know how to use medicinal plants—and for this reason they do not, along with the apparent ease of using biomedical remedies. She points out that the local clinic is actually just a dispensary of standard medications for diabetes or vaccinations for children. Thus, when the medicines are cheap and easily accessed, people she knows are more likely to opt for the biomedicine. Also, other people use modern medicine because they think it is more powerful—the remedies are quicker. But most people "just don't have faith in medicinal plants anymore."

Six months after our interview, I received an update on Lourdes and her condition. She was back to work at the indoor plantation. The working conditions there are deplorable; in one week, a family of five with three of the family members working there takes home a total of about 2,000 Mexican National Pesos a week—just barely enough to get by. And, in spite of all her best intentions to avoid exposure to chemicals, she must do what she must do to sustain herself and support her family.

A few blocks away, I chat with a fifty-year-old homemaker, Claudia, who tells me she has always used medicinal plants. She explains that she uses medicinal plants because they are natural and because she likes to. Also, Claudia does not like the chemicals in modern medicine. Pharmaceuticals are full of chemicals and chemicals are unnatural and therefore bad for the body. But the main benefit for Claudia gains from using medicinal plants is saving money because biomedicine is expensive. Also, she comments that through her knowledge of medicinal plants she is able to cure her family, particularly her grandchildren.

Claudia learned the cures she knows from her grandmother. However, Claudia no longer learns new techniques or plants. Sometimes she asks her neighbor for help. Her neighbor's deceased husband was a curandero, so "she knows a lot." Claudia says that others don't use medicinal plants because they simply do not know how to do so. Claudia observes about her fellow community members in Zacualpan that they are "all modern now" and go to the clinic or the pharmacy in Colima since they prefer modern medicine.

Heeding Claudia's advice, I head next door and speak with Celestina, a sixty-three-year-old homemaker. When Celestina was growing up, and later as she began to raise her own family, her family did not have enough money to go to the clinic if someone became ill. Instead, they used medicinal plants to cure themselves. It has been so long that now she is simply used to it, but before it was out of necessity. Interestingly, Celestina offers the same complaint about the chemicals in biomedical pharmaceuticals that are bad or toxic. Celestina feels she benefits by saving money healing herself and her family.

She learned about medicinal plants from her grandmother and other women in the community while growing up. In addition, Celestina consults her husband's big books to continue learning about new treatments and plant uses. He was a traditional healer and had bought a homeopathy book, among others, and he used those to teach her new things. Even though he has passed away, she still has those books and uses them to learn new medicinal plants and cures.

Celestina thinks that people don't use medicinal plants because they do not know about the plants and they are lazy. It is so much easier to just go to the clinic—especially since the clinic is so cheap (and often free, now). People think that modern medicines are "better, stronger, and quicker."

Meanwhile, I was also interested in speaking with biomedical practitioners, of which there is only one in Zacualpan. Pepe is a recent graduate of medical school, and at twenty-three years of age he was sent to Zacualpan to complete his social service requirement. Originally from Guanajuato, he says that he never really recommends the use of medicinal plants for cures because the people come to him for modern medical pharmaceutics—usually for their diabetes medicine. Pepe is not confident in plants and expresses concern about the lack of reliable science regarding possible effects and interactions with other pharmaceuticals.

When pressed, he confesses that in his own life he has drunk teas or simple home remedies his mother or grandmother gave to him. Other than that, he does not use medicinal plant remedies. The little that he has learned about simple home remedies and medicinal plants he learned from his mother, aunts, and grandmother. Pepe cannot really see any personal benefit to using medicinal plants. When he is sick he likes to use modern medicine. "It's better and more reliable."

Pepe thinks that people in the community do use medicinal plants, but that they should not. Why? "Because most people do not really have reliable knowledge on how to use medicinal plants, especially in terms of identifying plants correctly." Furthermore, modern biomedicine has not tested these plants for possible interactions with modern medicines. Regardless, people usually come to the clinic for "free medicines that I distribute, they do not use medicinal plants very much." His duty is to staff the clinic and dispense vaccinations and diabetes medications.

Pepe stresses his concern that people should not be using medicinal plants to cure themselves—"it is dangerous. They could get really sick because a lot of plants are poisonous." The only acceptable medicinal plant use he would recommend are simple home remedies such as teas.

There is a clear contrast in the terms and perceptions explained by the women and the biomedical doctor in Zacualpan. The women described biomedicine as pills that were laden with chemicals, but were strong and powerful—acting quickly and often being free and easy to access. In contrast, Pepe saw medicinal plants as part of the unknown, and therefore something that was dangerous. Indeed, the women tended to agree that many people no longer have the knowledge of how to use medicinal plants. However, they felt that this was unfortunate, and invested in seminars and books to learn more about medicinal plants. Traditionally, knowledge is learned directly from your forbearers, whereas learning from books and designated experts, "teachers," is a feature of modern modes of knowledge transmission. All of the informants talked about money; Pepe noted that community members come to him for free medicines he distributes though national health promotion initiatives. Again, the women noted the high costs of pharmaceuticals and the affordability of medicinal plants by comparison.

Ayotitlán

The traditional healers and individuals knowledgeable about medicinal plants are less easily identified in Ayotitlán. One practicing curandero, Consuelo, is sixty-five years old and lives about 200 meters from the Casa de Medicina Tradicional (discussed in chapter 9) in the center of the community. Consuelo says that he uses medicinal plants because they are good for curing and they are better (more effective) than modern medicine. As we began the interview, a woman showed up with her young male child who appeared to be about two years old. The child was crying and was diagnosed with *susto* (fright), bronchitis, and fever. He was taken to the front room which is used as a therapy and consultation room by Consuelo, who works from home. This is the only practitioner, and only community, with material culture related to medicinal plant knowledge in the Reserve area. (The Casa de Medicina Tradicional and Consuelo's consultation room.) In his consultation room, Consuelo massaged pomade on the boy's chest, calves, and feet. The ill child was also given a sweetened tea which his mother fed to him with a spoon. He calmed shortly after the massage and pomade were administered and fell asleep in his mother's lap. Though the child still had a slight fever, he was resting now.

Resuming the interview, Consuelo points out that he makes a modest living from curing with medicinal plants. He is careful to emphasize that he charges the bare minimum, accepting payments in kind. Consuelo complains about traditional healers in nearby communities who charge excessively (because they can; in more removed communities there is no access to emergency medical services). Consuelo charges very little because the people don't have much money and "someone has to cure them." Consuelo's mother was a *curandera* and she made him learn the trade. But he no longer learns about new plants or techniques, focusing instead on his specialty areas. Similar to Lourdes in Zacualpan, Consuelo has attended seminars and workshops in other parts of Mexico: Puebla, Veracruz, and Guerrero. Consuelo asserts that "in those regions there is actually more knowledge of plants. The plants are dying and disappearing in Ayotitlán." He buys dried herbs at the market in Autlán. Consuelo can see very few reasons why other people do not use medicinal plants, but suggests it could be because of "Fear. Perhaps people do not know how to use medicinal plants or have had a bad experience or have heard rumors of failed experiences. Or, they are rich," he tells me.

Consuelo's daughter, and assistant at mixing many of the teas and pomades, tells the story of how she recently gave birth to her first child. She went to Autlán in advance of her expected due date in order to be able to give birth in the hospital. She did this because she had access to the hospital and did not want to take risks. It is not that she is skeptical of traditional medicine, she asserts, but rather that she wanted to take advantage of her access to modern biomedicine.

Although she was firm that she has full confidence in traditional medicine, she continued on to share the story of a midwife in Ayotitlán who didn't get the placenta out and the woman died. This happened about five years ago, and even though the incident occurred in Ayotitlán and everyone knows about it, some women still go to her when they are giving birth.

Before parting, I wanted Consuelo and his daughter to suggest other medicinal plant experts in Ayotitlán. They told me that there are "at least ten healers, even more in the smaller communities." There is one curandera in Ayotitlán who "drinks too much." She is described as an alcoholic, but as someone who nonetheless knows how to heal using plants. Their stories quickly evolve into a discussion of the scandal that occurred regarding the financing of the *Casa de Medicina Tradicional* (Traditional Medicine House). After those interviews, I was never able to complete another full interview with other traditional healers or medicinal plant experts in Ayotitlán. When I approached the individuals I was referred to, they declined being interviewed (usually citing the expertise of, and sending me to, another unsuspecting neighbor).

Pepe, the medical doctor in Zacualpan, seemed like a moderate in comparison to his counterpart whom I met in Ayotitlán. The medical doctor stationed in Ayotitlán, Rodolfo, was also twenty-three and completing his social service. He did not see any use for medicinal plants, and seemed convinced that the local community members were not inclined to use them, mostly because they no longer possess much traditional knowledge. He said that traditional culture in Ayotitlán has been completely eroded and the community needs more exposure to modernity in order to learn to appreciate modern technology more. He sees the governmental program OPORTUNIDADES (a specific part of the PROGRESA program) as the primary culprit in creating a welfare-dependent community. He commented that people in Ayotitlán were "too lazy, simply having more and more babies and waiting for their government stipends each month."

Seeing no benefits of medicinal plant use for himself or for others, Rodolfo reports never using medicinal plants himself. He is more confident in modern medicine because "it is based on science and is simply better." He feels that "it is good that he is there to provide medicines for people and to help them learn more about modern medicine." He is looking forward to finishing his social service commitment so that he can go to the University of Guadalajara Medical School where he will specialize in obstetrics and gynecology. He has performed cesarean sections before and doing them excited him greatly, thus his choice in specialty. I cannot help but wonder if his perception of the economic forces that drive Ayotitlán are not also a factor in his lucrative choice of specialty. This young medical doctor may not represent the perceptions of all those who staff the clinic in Ayotitlán regularly (five nurses and two doctors).

A male nurse who works full-time at the clinic in Ayotitlán, Efren, also completed an interview with me. He is thirty-three years old and originally from a nearby community. Now he and his wife live in Ayotitlán and have a young toddler. He says that it is good that community members use medicinal plants. The people value medicinal plants, have a great deal of knowledge about their uses, and have a certain amount of pride in their medicinal plant repertoire. In fact, Efren says that he has used medicinal plants himself, primarily when his grandmother has taken him to the traditional healer or has brought things to him. His grandmother is also the person from whom he learned about medicinal plants. He personally benefits from the use of medicinal plants because they are "very healthy and effective in promoting health."

When asked if he thinks many community members use medicinal plants he was baffled by the question. "Of course they use medicinal plants—they are an indigenous community!" The people there still know a lot about plants and that is how they are used to curing themselves and it is their preferred mode of curing. Yes, people in Ayotitlán also come to the clinic, but usually after the person is really sick or only if they don't know what is wrong. However, they can usually cure themselves using plants. Additional benefits Efren sees that come from using medicinal plants is their accessibility for the community members and the capitalization on cultural knowledge. There really are no disadvantages to using medicinal plants, according to Efren. It is part of the local culture and identity.

An unexpected interview started itself one evening in Cuzalapa, about one hour from Ayotitlán, where I chatted with a thirty-four-year-old woman who works in the local artisanal coffee producing cooperative. Luz is somewhat of a traditional healer, although I came to find this out in a roundabout way. Amidst her plentiful gardens there are any number of flowers, trees, plants, vegetables, herbs, and fruits. The amazing part is that she knows a use for each one (and she points out that she just learned that the leaves of a tree which she thought had no use can be wrapped around the feet to help with calluses). Luz enjoys giving tours of the area where she lives and works—it seems she is always pointing out some plant and its usefulness. I went with her to feed the chickens and the puppies after the sun had gone down. I asked her why she knows so much about medicinal plants and she was eager to share. She was raised by her grandmother in a different community in the Reserve where she was taught about plants and trees. She explained that when she was growing up everyone had to use traditional medicine because there was no clinic or hospital that was accessible. Whenever anyone got sick, they "cooked something up." She said that now other people in Cuzalapa still use traditional medicine—not just her, but some of the other ladies from the group that work in the coffee cooperative together. These women who use medicinal plants share and exchange plants with each other since they each have different plants growing in their individual household gardens. Luz tells the story of one day when a woman came and apologized to her, saying that the night before she had cut some of one of her plants. Luz smiled and assured the woman it was not at all a problem—"that's what the plants are there for: to help people when they needed them."

Luz went to her office and found a report she had completed with a former student regarding medicinal uses of the tomato (principally) but also any other food plants. They interviewed people in the community—many of whom have now died—and asked them how they prepared different home remedies. This is one of the few examples of material culture of medicinal plants in the Reserve. They prepared a report of all these uses as part of the way to retain that cultural knowledge. There are a few print copies of the report that circulate in the local community and CUCSUR. It cannot be certain what impact the report had in Cuzalapa, except to note that it was enthusiastically undertaken with the involvement of many elders in the community (many of whom have since passed away).

I asked Luz why some people did not use traditional medicine anymore. She said because now it is easier (and for some, more convenient and comfortable) to simply go to the doctor and take a pill. But, she noted, these medicines are strong and the people are using them without really understanding them. Then she pulled out an announcement sent by the government in the mid-1990s that was addressed to all doctors, healthcare workers, and community leaders. To paraphrase, this memorandum explicitly stated that, in accordance with the

Salud Para Todos (Health for All) 1995-2000 Program, no one should be using traditional medicine or home remedies whose interactions and secondary effects were not studied. Luz was angered by this letter. For her, it was proof of how "the government is taking away the peoples' cultures and customs."

Lessons Learned from the Stories Told

These stories highlight important aspects of using plants to promote and sustain health in the Reserve. The most commonly mentioned or implied factors include: money; distance to modern medical services; cultural value; availability of plants; and chemical characteristics of modern pharmaceutical drugs in contrast to non-chemical plants. Knowledge of medicinal plants is, logically, associated with availability of the plants, availability of alternative health services, and accessibility of both modern pharmaceutical drugs and medicinal plants. Pushing beyond these rational factors of medicinal plant knowledge and use, the moral meanings of plants are exemplified in the stories told. As these case studies showed, individuals use medicinal plants to fulfill moral needs—because of the cultural value of medicinal plants—and thus the understanding of medicinal plant knowledge is more than a rational model of economic livelihoods.

The people in the communities navigate a medically plural society using both substantive and formal rationalities. Though they value their cultural heritage, they also recognize the costs and potential benefits of modern medicine—it is both powerful and expensive, thus it is used when available. In the interview, money saving was the most commonly mentioned reason for using medicinal plants. From the perspectives of traditional healers and plant experts, one of the principal benefits of using medicinal plants is saving money—or the lack of money to buy modern medicines is a reason to turn to medicinal plants. Health clinics are often hard to reach, and frequently unstaffed or ill-equipped, in many communities, thus the reference to inaccessibility is not simply financial but logistical. Indeed, even the biomedical practitioners who were interviewed made reference to welfare and social programs that make healthcare and modern medicine more accessible to community members. The financial aspects of affordability and accessibility should not overshadow other clear benefits that were mentioned by the interviewees.

"Chemicals are bad," was the next most common response. Indeed, all interviewees except one cited *quimicos* (chemicals) as the bad thing to be avoided, hence the reason for using medicinal plants, and an argument for why it is better to use plants to heal infirmities. What is a chemical? According to the interviewees, a chemical is defined as something that is not "natural." Plants are natural and therefore good, or at least better than chemicals. The differences between the MPK experts in Zacualpan and Ayotitlán (including nearby Cuzalapa) are demonstrated in their perceptions of biomedicine. The interviewees in Zacualpan were more interested in the growing acceptance of biomedicine and less concerned with MPK as a source of livelihood.

Other important trends and observations emerged from these interviews. Biomedicine and allopathic medicine are extensions of the government—with all clinics and clinicians being part of national programs to provide healthcare services to rural areas. Similar to the fieldworkers of the management of the Reserve, representatives of allopathic medicine are given to instructing and "correcting" locals, not so much as a form of education, but rather as part of the paternalistic model of taking care of locals and giving them guidelines that are "for their own good," as was demonstrated in interviews with the biomedical doctors and resonated in other conversations with representatives of the Management of the Reserve. In this way, the same political-economic processes driving the management of forest resources are influencing the availability and use of healthcare services. Doctors recommend that community members not use medicinal plants because they are at the risk of being harmed by the use of MPK. The irony is that the primary causes of the illnesses being treated by the biomedical health services (respiratory infections, diabetes, and hypertension) are byproducts of the modernization and development of their communities. Exposure to modernized diets and toxic chemicals in the agribusinesses where many work are sources of the illnesses. Cultural heritage and knowledge are devalued by the biomedical doctors, except for the nurse in Ayotitlán who saw that his fellow community members use medicinal plants with confidence and safely. The nurse (who was also originally from Ayotitlán) embraces his own identity as an indigenous person who uses medicinal plants; he represents an intersection of the two ethnomedical systems.

There are other reasons to expect the regulation of forest resource usage to impact the knowledge and use of medicinal plants. The extraction of medicinal plant knowledge by outside agents or institutions is a number one concern; the ability to learn about medicinal plants and their uses often comes from outside sources, as well. The primary place for continuing education about medicinal plants is in larger urban areas and at seminars or workshops. The diffusion and cultural borrowing demonstrated in the sharing of MPK across communities in Mexico can lead to an important conclusion: the nature of intracultural variation of MPK is such that it may not be localized enough to be considered part of a forest conservation paradigm for Mexican forests. A traditional agricultural ecosystem is classically conceived of as a localized system of cultivation with accompanying specialized knowledge about environmental resources. If the knowledge of specific medicinal plants is not necessarily connected to a geographic location, then the motivations for conservation of traditional agricultural ecosystems are weakened. The quantitative data I collected, presented next in chapter 8, also seem to support this conclusion.

Chapter 8
Medicinal Plant Knowledge in the Reserve

In chapter 4, I provide a detailed overview of the field techniques commonly used for the study of TEK. In this chapter, I present the results of my study of a specific type of TEK—medicinal plant knowledge.[1] By focusing on medicinal plant knowledge (MPK), the data I present in this chapter is situated to describe how economic livelihood is associated with the amount of cultural knowledge of forest resources in a protected area. I constructed my exploratory study on previous research of the distribution of ethnobotanical knowledge, in these same study communities and also cross-culturally. I wondered if individuals who reside in households wherein the primary economic activities were based in subsistence agriculture would know more medicinal plants than individuals whose households relied on intensive animal husbandry for subsistence. I collected survey data that can describe some possible trends, illuminate areas for further inquiry, and suggest some preliminary comparisons with previous work in the Reserve (Benz et al. 2000) as well as prior research in other cultural settings that has been collected at the individual, household, and community levels.

Benz and his colleagues analyzed the relationships between knowledge of plant use and various indicators of community change and modernization (1994; 2000). In eight communities in the Reserve, Benz's research team used informal freelisting in interviews to identify the categories of useful plants. They also used herbarium voucher samples to ask informants about the ability to identify and name a plant, and also to provide any information about the uses of the plant. They used these data to create various community-level indices of ethnobotanical knowledge. The indices were then used to assess the relative importance of different plants across communities. The communities were ranked in terms of marginality based on national census data. Benz and his colleagues looked at the percentage of households in a community that had certain housing conditions (dirt floors, electricity, plumbing, sewer, or one room only) along with the percentages of literacy and primary school completion. Their results showed that adoption of modern services in some, but not all, communities does seem to correlate with a decrease in knowledge of local plants (Benz et al.

2000). The particular community of Telcruz (see below), however, confounded their analysis, because here there was no apparent relationship between the community's socioeconomic profile and the aggregate measure of plant knowledge for the community. They have suggested, and my findings concur, that further investigation into the economic activity profile of households and land usage may more adequately explain variation in traditional medicinal plant knowledge (Benz et al. 2000).

In the Reserve, individuals' livelihoods are related to the extent and nature of their interactions with the market economy. For my study, I approximated "market integration" as the distance to a regional market, as well as the intensity of agriculture and animal husbandry, following other studies that operationalized market integration (Godoy 2001; Reyes-Garcia 2001). Because the Reserve is a protected area, the livelihoods of the indigenous people who live there are complicated by restrictions on natural resource uses (IMECBIO 2000; Jardel 1992). Economic changes (particularly increased integration with markets outside the community) are related to other processes like "modernization," but not only to measures of modernization or "development." All economic change does not equal modernization nor changes in traditional ecological knowledge (Berger and Luckman 1966). However, consumer products and services, labor patterns, and worldview are interrelated. Exposure to foreign-made consumer products and access to new technologies are examples of the pervasiveness of outside markets in peri-urban and rural Mexico. Employment, purchases, and relationships with others within the community are all related to the broader structures and relationships between individuals and institutions. Individuals and households increase their stake in the market economy as they transition from relying on trade and self-sufficiency to reliance on market interactions and products. Economic changes are central components of the ways in which individuals and communities are facing new problems and opportunities in contemporary Mexico.

Anthropologists are interested in intracultural variation in both general and specialized knowledge (Boster 1986). In the past, anthropologists defined culture as a system of shared knowledge and developed cultural consensus as an effective method for modeling shared cultural knowledge or "cultural domains" (Romney et al. 1986). A cultural domain is a body of knowledge that is determined and organized by culturally shared classifications, symbols, and meanings (Berlin 1992). Cultural knowledge about the environment informs the ways in which individuals interact with their environments and is passed on through both formalized and informal knowledge transmission systems—schools, books, experiences, elders, and peers (for a discussion of culture learning in indigenous societies see, for example, Garro 1986). Cultural domains represent specialized knowledge, which has a patterned distribution (Boster 1986). Those intracultural patterns of distribution are influenced by various factors; my research explores one specific area of specialized cultural knowledge (medicinal plants) that is a locally salient category (Benz et al. 2004). I measured the amount of medicinal plants an individual could list as part of my research study to help advance our understanding of variation in MPK.

I limited my investigation of medicinal plant knowledge to botanical products and did not seek to identify what people actually do with these plants nor to report their capacity for using medicinal plants, which would include practical and skills dimensions (Reyes-Garcia et al. 2007). Although medicinal plant knowledge might be highly specialized in some groups, previous research in the Reserve had demonstrated that "medicinal" was the most commonly reported use of forest plant resources (Benz et al. 1994). The *a priori* relevance and ubiquity of the cultural domain "medicinal plant" means that the category of medicinal plant was a locally valid area of inquiry. I could reasonably expect the knowledge of medicinal plants to be widely distributed within the cultural group, both amongst specialists and non-specialists (Romney et al. 1986).

I employed freelisting and household surveys to describe and explore possible relationships between social and economic characteristics and medicinal plant knowledge (Godoy 2001; Romney et al. 1986). In this study, I explored the connections between the most commonly reported economic and subsistence activities and differences in medicinal plant knowledge in three communities within the Reserve. I expected that communities in which the households tended to rely more on non-farming activities (especially intensive cattle ranching) would list fewer medicinal plants than individuals living in communities where most households relied on subsistence farming. I suspected that there might be a pattern related to age or gender and the amount of medicinal plants an individual could list, insomuch as these variables might also be related to differences in the socioeconomic profile of individuals in the study communities.

Freelisting is used by ethnographers to identify a cultural domain (Thompson and Juan 2006). The number of items in that domain can be used as an approximation for the amount of a specific type of cultural knowledge that a person holds. Again, the research I present here does not address traditional medicinal knowledge beyond the knowledge of plants—specific remedies or preparations were not recorded. (Working with TEK among indigenous peoples necessitates a high degree of sensitivity to the issues of intellectual property rights and benefits of the research to the community [Berlin and Berlin 2005].) In order to carry out my research, I went through a series of discussions with local community leaders and arrived at an agreement (which I detailed in chapter 1). The permissions I received to collect data regarding medicinal plant knowledge restricted me from reporting actual plant names. I was not interested in specialist knowledge (such as actual remedies or treatments) but instead in how much cultural knowledge individuals in different communities possessed. Thus, it was not at all a hindrance to the quality of the data that I collected to code reported plant names and report neither plant names nor indigenous healing practices.

The medicinal plant knowledge was an individual-level variable, but the livelihood information was reported at the household level. I used a purposeful sampling technique to obtain a stratified sample representing different ages and genders in the three study communities, and I also used purposeful sampling for the qualitative data collection. Data for the household variables were collected from the main study sample, which was divided into three cohorts to obtain a sample of adult men and women in the communities of Ayotitlán, Telcruz, and Zacualpan (shown in table 2.2). I sampled nearly 14 percent of households in Zacualpan, 15 percent in Telcruz, and 23 percent in Ayotitlán, including just over 3 percent of all adults in each community in my sample (since only one adult per household was included). I only interviewed one individual from each household, and he or she was asked to report household-level information regarding livelihood and economic activities.

The community Telcruz is particularly interesting because of the overall socioeconomic profile of the community: Telcruz is remote geographically, but has higher indicators of general social and economic achievement than Ayotitlán. In terms of the social and economic activity profile measures collected for my investigation, and consistent with secondary data reports such as the national census, I found that Telcruz is neither as marginalized as Ayotitlán, nor as "modernized" as Zacualpan.

Creation of Indices

Using the extensive data I collected, I compiled a rating of individual expertise in medicinal plants compared to others in the community based on list length. Freelisting allows me to describe the cultural domain of knowledge and compare community members' list lengths as a type of proxy measure for relative amounts of knowledge held.

I aggregated some of the data in order to facilitate comparisons across individuals and households.[2] The calculation of these indices was linear and provides for a comparison across individuals and households in this study. In this way, the indices which are summarized in table 8.1 act as proxies for economic activity and socioeconomic status (measured at the level of the household). I calculated the indices so that a higher index score indicates: greater degree of contact with communities outside the Reserve; higher level of education; higher standard of living in the household; greater total cash income; heavier reliance on technology; and more intensive agricultural techniques. None of these indices should be seen as a stand-alone measure of socioeconomic status or degree of relationship with the outside world. Taken together, these different scores can help shape a picture of the types of relationships and important variables associated with knowledge of medicinal plants and degree of modernization.

Table 8.1. Construction of socioeconomic indices (yes=1, no=0).

Variable Name	*Questionnaire Item*
Demographic & Household	**Higher score reflects higher degree of contact with communities outside the Reserve and higher level of education. Higher score reflects a higher standard of living in the household.**
	Higher score reflects higher degree of contact with communities outside the Reserve and higher level of education. Higher score reflects a higher standard of living in the household.
	Subject not born in same community where living now.
	Subject has lived outside community for more than 30 days.
	Roof material not cardboard or plastic.
	Non-dirt floors.
	Have plumbing.
	Have running water.
	Have hot water.
	Do they have a household garden?
	Have more than 5 chickens.
	Have more than 5 pigs.
Economic	**Higher score indicates higher overall cash income.**
	Higher score indicates higher overall cash income.
	Have they sold anything in the two weeks prior to interview?
	Have they traded anything in the two weeks prior to interview?
	Have they worked for a wage in the two weeks prior to interview?
	Have they received any remittances or government stipends in the month prior to interview?
	Have they borrowed money from anyone in the prior month?
	Have they lent money to anyone in the prior month?
Agricultural	**Higher score indicates heavier reliance on technology and more intensive agricultural techniques; lower score indicates a relatively smaller level of production.**
	Higher score indicates heavier reliance on technology and more intensive agricultural techniques; lower score indicates a relatively smaller level of production.
	Did they use a tractor or machine to plow and/or plant?
	Did they fumigate?
	The quality of corn this year was either bad (0 points), regular (1 points), or very good (2 points).
	Do they plant every season?

Description of Variation in Medicinal Plant Knowledge

Knowledge of medicinal plants is the dependent variable and was measured at the individual level as the length of the list. I measured the other indices at the level of the household in which the individual participates and resides. I looked to these data to describe trends in the amount of medicinal plants listed and the other socioeconomic and demographic indices.

The most plants listed by a single informant were twenty-three—a number reached only by individuals in Ayotitlán and Telcruz. Zacualpan was the only community in which some informants (7) could not list a single medicinal plant; but everyone who could list at least one plant listed two or more. It is not possible to compare the list length with the data previously collected by others regarding medicinal plants in the Reserve because of the nature of the question asked. Whereas Benz and his colleagues had used herbarium voucher specimens to elicit information about usefulness of a plant, I did not prompt the study participants with any plant samples. It is possible that using plant samples to elicit responses would have resulted in much longer reported list lengths. Some individuals may have perceived the filling of a single column on the page as a full response, and thus may have artificially truncated their response. By repeating the probing question three times during the freelist interview, I attempted to avoid this possible flaw in the data I collected.

Table 8.2. Plants listed by community and whole sample.

	Zacualpan	Telcruz	Ayotitlán	Total
Count of unique plants listed	116	78	120	215
Count of unique plants with at least three mentions	25	18	38	72
Mode list length	6	3	9	4
Median list length	5	5	9	6
Mean list length	5.70	6.58	10.18	7.15
Standard deviation of mean list length	3.32	4.21	5.85	4.72

There is no significant difference between the list lengths of men and women study participants, according to the Mann-Whitney test ($z= -0.097$, $p=.923$) where the total number of plants listed is not assumed to be normally distributed (Leeper 2008). Therefore, I found that gender alone is not a good predictor of an individual's amount of medicinal plant knowledge. The results, summarized in table 8.2, show that the average list length is similar for Zacualpan (5.70) and Telcruz (6.58) though noticeably higher for Ayotitlán (10.18). The median list length of medicinal plants was the same for both Zacualpan and Telcruz (5), while the median was longer in Ayotitlán (9). The mode shows a different trend, wherein Ayotitlán has again a modal list length of nine, but Telcruz has a much smaller modal list length (3) than Zacualpan (6). In Telcruz, the number of unique plants that were listed (78) was lower than in either Zacualpan (116) or Ayotitlán (120).

Livelihoods and Socioeconomic Indices

Agriculture is much more central to the livelihoods of households in Ayotitlán, where the average number of plants listed by each informant was nearly double the means in Zacualpan and Telcruz. Farmers in Ayotitlán are more likely to have their own seeds (78.8 percent contrasted with the 15.2 percent who purchased all of the seeds they used). Also, as seen in table 3.1, in Ayotitlán the farmers plant more hectares than households in either Telcruz or Zacualpan, on average.

As illustrated in table 3.1, the average field size is greatest in Ayotitlán (1.88 hectares) and smallest in Zacualpan (1.24 hectares), while Telcruz lies between the two (1.58 hectares). Milpa fields are mostly found in Zacualpan (22 percent) and Ayotitlán (18 percent) but in all communities, corn is the most common crop: Zacualpan (43 percent), Telcruz (89 percent), and Ayotitlán (76 percent).

Households in Telcruz tended to either already have (55 percent) their seeds, or purchase seeds (34 percent). In neither of the other two communities are seeds purchased as frequently: 17 percent in Zacualpan and 15 percent in Ayotitlán. (The only major exception to the importance of where seeds are obtained is evidenced by the 33 percent in Zacualpan who did not need seeds—either because they did not perform any agriculture or because they were harvesting perennial plums off of trees.) Of course, it is most common to use seeds retained from the previous crops (35 percent in Zacualpan did this, and 79 percent in Ayotitlán).

Table 3.1 also summarizes the types of cattle and milk cows held in the three study communities. Very few households in Ayotitlán have any cows (9 percent) and the average total number of cattle and cows held is 1.24. In Zacualpan, on the other hand, an average of 4.46 total cows and cattle are held, but of course these are held by a third (32 percent) of the study sample. Only 11 percent of households surveyed in Telcruz have any cows and the average total there is comparable to that of Ayotitlán (1.34).

Identifying Relationships between the Indices

The data I collected suggest that individuals who live in communities where, on average, there is a heavier reliance on agriculture, and not intensive cattle ranching, for their livelihoods listed more plants. Zacualpan has the most cattle and cows per household (nearly three times the average amounts held by households in Ayotitlán or Telcruz); individuals in Zacualpan also had much shorter lists of medicinal plants on average.[3] The higher reliance on animal husbandry (i.e., intensive cattle ranching) as a source of livelihood appears to be associated with shorter list length. The type of animal husbandry and the extent of grazing will be a factor in what interaction the herders have with forest resources. Intensive cattle ranching in the Reserve and Mexico is related to clearing forest and fallow areas for grazing and the sale of cattle for cash, rather than for personal consumption (Gerritsen 2002; IMECBIO 2000).

In my research, I was looking for logical relationships between my explanatory, socioeconomic data, and the measures of medicinal plant knowledge. My research was informed by prior studies in the Reserve (Benz et al. 2000; Benz et al. 1994) and other areas, which have looked at the roles of gender, age, socioeconomic status, modernization, and household characteristics as possible explanations of intracultural variation in medicinal plant knowledge (Godoy 2001; Reyes-Garcia 2001).

There have been sweeping modernization programs and introductions of advanced technology to the Reserve since the 1990s, yet the loss of traditional knowledge has not been uniform across these modernizing communities (Benz et al. 2000). The fact that traditional medicinal plant knowledge does not adhere to a linear model—such that increasing modernization leads to a decrease in traditional knowledge—is significant. There are many aspects of cultural, social, ecological, and economic interactions that may shape the distribution of specialized cultural knowledge. Economic activities that require individuals to interact more with their environments may be related to having more TEK, and individuals may have an incentive to increase and develop TEK to provide products from the environment to the market (Ahmed et al. 2010). The many determinants of TEK are most likely interrelated differently in different cultural groups.

Among studies of traditional forest resource knowledge, there continues to be a large gap in comparable research studies. There is little ethnobotanical literature on cultural consensus and community knowledge. In addition, few existing studies include samples of more than fifty informants (Reyes-Garcia et al. 2007). Studies very rarely report their informant counts and other data essential for comparison across studies. Of those studies that do exist, individual-level variation is usually not reported; aggregate community level data is the preference. I collected individual- and household-level measures of TEK and economic activities and socioeconomic status, but likewise found it was necessary to aggregate the data at a community level to see apparent trends. The household census data I collected provide descriptive and summary measures; these data suggest that there may be significant relationships between economic activity, socioeconomic status and plant list length. It is my hope that the rigorous individual- and household-level data I collected may contribute to future and ongoing research in these same study communities.

Other Possible Explanations of the Findings

There are some additional avenues to be explored in understanding the variations in medicinal plant knowledge in the Reserve. Any discussion of the role of outside markets and economic activities in Mexico cannot ignore the influences of national and transnational migration. Though my investigation did not deal directly with migration, it is an important domain that overlaps with knowledge and knowledge transmission. Whether the individual participant has lived or worked in another place, or someone is sending remittances from another location, the role of migratory work in the Reserve needs to be considered. We know that cultural knowledge is impacted in its diffusion and dimensions by migration patterns (Berkes 1999).

Recall from chapter 4 that some studies have contended that when indigenous knowledge seems endangered, it should increase in local value because it is "rare" and thus more valuable (Benz et al. 2000; De Walt 1994; Godoy 1994). It is sometimes suggested that where TEK is very specialized and not widely shared, but has potential market value, individuals will proliferate that domain of cultural knowledge (Ahmed et al. 2010). The most marginalized community in the Biosphere Reserve, with the lowest levels of wealth and socioeconomic status was Ayotitlán; its households also had the longest average list lengths. Ayotitlán has been characterized as an indigenous community with a relatively high degree of TEK (Benz et al. 2000) and where cultural knowledge and heritage is reportedly highly valued (Olson 2009). Ayotitlán has demonstrated resistance to the presence of governmental—or other institutional—programs aimed at promoting sustainable development (Olson and Gerritsen 2011). The purposeful resistance community members demonstrate to outsiders and outsider programming indicates that institutional promotion of conservation and conservation-oriented programs is not necessarily a requirement for conservation of traditional knowledge. In the Reserve, other research teams have used structured interviewing techniques to identify useful plant species in an effort to inform sustainable development strategies. They ask the same question first posed by Foster in 1962: To what extent can traditional knowledge persist in modernizing societies? (Benz et al. 2000). In particular, we still do not have a clear picture of which cultural domains these changes occur in most prominently.

Various researchers have argued that local indigenous knowledge of forest resources is critical to sustainable resource use and conservation (Buenz 2005; Caniago and Siebert 1998; Cox 2000; Icamina 1993; McNeeley et al. 1990; Prance 2000). Indigenous technology is often associated with sustainability and lower negative impact on the environment (Buenz 2005; Caniago and Siebert 1998; De Walt 1994; Gerritsen 2002; Oldfield and Alcorn 1987). In developing countries, interests conflict between desires to encourage indigenous language and knowledge to promote better economic and social opportunities, versus promoting advanced technology and modernization introduced largely by governmental agencies (Brown and Wyckoff-Baird 1995). The amount of TEK that individuals hold may help us to understand the types of cultural knowledge that remain important in indigenous communities where sustainable development and conservation programs are underway. Are there certain types of cultural knowledge that persist and are transmitted continually across generations, in spite of broader socioeconomic changes?

The social-political processes shaping the landscapes in the Reserve are also influences on the ethnobotanical knowledge held by community members (Olson 2009). Paternalistic models of resource management dominate in the Reserve: residents are advised what types of agriculture to carry out and where; they are also instructed on how to use (or not use) forest resources such as medicinal plants (Gerritsen 2002; Jantzi 2000; Jardel 1992; Olson 2009). There has been some contention in conservation literature as to whether or not local actors are appropriate deciders of resource management (Wilshusen et al. 2002); yet case studies have asserted the importance of local decision-making and empowerment (Brechin et al. 2002). Local-level processes within communities that shape interactions between humans and environments are likely to impact the nature and distribution of ethnobotanical knowledge held by community members. Specialized knowledge, like medicinal plant knowledge, has a patterned distribution that is influenced by economic, social, and other individual characteristics. In a recent study, Ahmed and colleagues (2010) found that expansion of specialized knowledge of certain types of agroforested teas was related to product demand in international markets. A common conservation paradigm suggests that people need to "know" a species or ecosystem in order to conserve it; anthropogenic relationships with environments are purposeful and intended to meet myriad subsistence, aesthetic, or other goals. At the time of writing, concerns regarding neoliberal environmental policies are challenging the simple notion that placing market values on natural resources will augment biodiversity conservation (Igoe and Brockington 2007; Fletcher 2010).

Conclusion

The increasing interconnectedness of local markets across national boundaries is an unfolding process accompanied by transmissions of knowledge and value systems (Ahmed et al. 2010). Globalization contributes to the interconnectedness of different cultures, and contemporary environmental issues are understood and commonly addressed on a global scale (Alcorn 2001; Gerritsen 2002). My purpose in this chapter is not to flesh out the ways in which neoliberal capitalism as a global phenomenon influences approaches to conservation (see Igoe and Brockington 2007; Fletcher 2010); nevertheless, the significance of these data should be understood within the broader context of neoliberal conservation and development strategies. Patterns in the distribution of cultural knowledge, such as MPK, can shed light on some of the ways in which globalization relates to indigenous cultures (Godoy et al. 2005).

My research has some important implications for ethnobotanical studies in the Reserve, as well as other cultural areas. Efforts can be made to preserve cultural knowledge of medicinal plants and ensure that this specialized knowledge is not completely displaced. Community members, farmers, and ranchers are interacting with their natural environment every day; the knowledge of forest resources is a dynamic system of learning, evolving, and adapting (Gerritsen 2002). The biodiversity that we seek to conserve in the Reserve is not static and is directly intertwined with the traditional knowledge systems associated with resource uses (Gerritsen 2002; van der Ploeg 2008). Gerritsen suggests that we should not only attempt to conserve existing traditional knowledge systems, but also to recover some of the cultural knowledge that has been lost due to increasing globalization and reliance on modern technologies (2002, *see also* van der Ploeg 2008). Certainly, many social scientists advocate for the conservation of traditional knowledge as part of the integrated conservation and sustainable development of protected areas (Benz et al. 2000; Brechin et al. 2002; Buenz 2005; Gerritsen 2002; Godoy 2001; Oldfield and Alcorn 1987; van der Ploeg 2008; Wilshusen et al. 2002). The question remains how to do so in an effective and socially just manner. Cultural heritage is part of human capital that is valued and sought to be included in integrated conservation development programs.

Notes

1. Portions of this chapter are based on my review article, "Anthropology and Traditional Ecological Knowledge: A summary of quantitative approaches to traditional knowledge, market participation, and conservation" which was published in 2013 in *Culture, Agriculture, Food and Environment* 35(2):140-151.

2. I used the dichotomous (yes/no) variables to create a set of socioeconomic indices for different household level attributes. I constructed these socioeconomic indices by combining dichotomous variables with a few constructed variables that make sense from a common sense standpoint.

3. The possible relationship between subsistence and economic activities and knowledge of medicinal plants was only borne out at a community level.

Chapter 9
Final Thoughts

By way of concluding this ethnography, I share one salient story of traditional medicinal knowledge and development programming in the Reserve. An example of a "win-win" integrated conservation and development program highlights the important themes that have emerged throughout this ethnography of indigenous knowledge and development. The example demonstrates that in fact people living in the Reserve are making trade-offs and hard choices each day and in regards to development projects. I share this example to underscore the important lessons learned in this ethnography about indigenous people, traditional ecological knowledge, conservation, and development in Mexico.

La Casa de Medicina Tradicional

Throughout my time in the Reserve, I was frequently told about the Traditional Medicine House which is located in the central plaza, adjacent to the local Catholic Church in the small mountain community of Ayotitlán (the most traditional of my study communities). The place was constructed of brick, plaster, and a metal roof (much nicer than any of the domestic residences in Ayotitlán which usually have dirt floors). But of course, this was not a private residence. On the front there is a worn out inscription that reads: *Casa de Medicina Tradicional de Ayotitlán* (which means "The Traditional Medicine House of Ayotitlán," which I refer to simply as La Casa).

La Casa had been constructed in 2002 and I first visited the location in 2007. By that time, the place was already vacated and in a dilapidated state. Within just five years of its construction La Casa had failed. I was consumed by the question of *why*—why had La Casa failed?

La Casa was constructed in the most marginalized community, Ayotitlán, with the goal of capitalizing on the rich cultural heritage and promoting health and well-being in the community. Ayotitlán has been described as the most "traditional," with the greatest amount of indigenous heritage, and a great deal of resistance to development programs and interventions. Ayotitlán is inhabited by indigenous people who are of Nahua descent and live mostly from their swidden farming and also remittances from family members who participate in migratory labor (both within Mexico and internationally).

It is in this context that the idea emerged to capitalize on the local cultural heritage and fill a gap in the need for health care in the community. There was an old rural health outpost that was about an hour walk from the main center of the village. The rural health outpost was not regularly staffed. Situations requiring immediate medical attention were usually attended by local curanderos or other individuals who were considered experts in medicinal plants or healing (including individuals who call themselves naturistas).

In order to capitalize on the local human capacities, and support the retention of valuable traditional ecological knowledge (this is, after all, on a Biosphere Reserve), the Management of the Reserve obtained a grant to construct and help maintain La Casa. La Casa had three examination rooms, a waiting room, full bathroom, and a kitchen—it was designed to serve all of the community healthcare needs. The yard's landscaping had space for gardens. The idea was that local curanderos would work at La Casa, staffing it, and over time build the gardens by transplanting foraged medicinal plants from the wild. Eventually, the gardens of La Casa would serve as a local pharmacy for the curanderos and supplant the need to forage for medicinal plants in the wild.

In a way, it was intended to be a "win-win" development program: promote the retention of traditional ecological knowledge which is vital to natural resource management, provide improved access to health care, promote local cultural heritage and traditions, and provide an income to traditional healers who staffed La Casa (McShane et al. 2011).

When I interviewed local people, they gave me two types of reasons for *why* La Casa had failed. First, since its inception there was corruption with the funding for La Casa (specifically, the embezzlement of federal grant monies that were earmarked for its maintenance and supplemental stipends for curanderos). Secondly, it was underused. The arrival of biomedicine and practitioners who discourage the use of traditional medicine was a reason that some people gave for not using La Casa.

I need to elaborate a bit more about the embezzlement, however, because it illustrates one of the ideological tensions between local indigenous people and the national government. The community representative who was the coordinator for the project was of indigenous background, and had lived in a nearby community for most of their life. However, this individual had left and gone to the university—studied botany, and came back to conduct community-based participatory research projects on ethnobotany. Coming back into Ayotitlán, this person was viewed as an outsider, despite their indigenous heritage, because they had become acculturated into the outside world. By attending college and taking up the work of the government and university, the allegiances and ideologies held by this individual were viewed as incongruent with community members'.

With regards to the second explanation of its failure, it just so happens that not long after La Casa was completed, a new biomedical clinic was also constructed a few blocks away—also near the center of Ayotitlán. This new clinic was staffed regularly by at least one biomedical doctor (a recent medical school graduate completing his or her social service) and two to three nurses. The biomedical physicians reportedly told community members that they could not treat them if they were using traditional medicine because "there could possibly be dangerous interactions."

By 2007, within five years, La Casa was abandoned, and being used as a storage locker for the local church. What I suggest is that La Casa's condition is symbolic, if not symptomatic, of community-based development projects in Ayotitlán.

I found that La Casa is an allegory that can help deconstruct the past attempts at a "win-win" community development strategy which might have been better framed as a set of "trade-offs and hard choices" (McShane et al. 2011). In this example, we can see how national-level rhetoric and policies are disjointed in regards to community development programs related to health and conservation. Through the example of La Casa, and the related experiences of traditional healers in Ayotitlán, it can be seen that local people actively resist what they perceive is an increasing domination of their traditional knowledge and culture. The tension between the national and local levels is played out in the failure of La Casa.

The Mexican model of a biosphere reserve is based on the principles of sustainable development and conservation by emphasizing community participation and empowerment of local people. The Mexican model is unique, perhaps, from some other natural resource protected areas, in that it is supposed to help retain cultural heritage and promote and community development. However, the rhetoric of national policies, and the local realities that people experience, can be incongruent. Additionally, since 1988, the health sector in Mexico has been engaged in a process of privatization of healthcare (Harrison 2001). As an indigenous community, this process may not resonate with traditional cultural values and medical systems (as I found in Ayotitlán).

As I discussed in chapter 5, stakeholders and key actors in the Reserve can be classified in four groups: (1) community members and institutions, (2) state-sponsored conservation and development agencies, (3) scientific research institutions, and (4) for-profit companies. Government agencies, research groups, and private industry can all be characterized as secondary stakeholders. From a local perspective, they are "outsiders" since they are not only foreign to the community, but are often perceived as having ideologies different from or even in conflict with those of community members.

Biomedicine in Mexico is an extension of the government—with all clinics and clinicians being part of national programs to provide healthcare services to rural areas. Similar to the fieldworkers of the management of the Reserve, representatives of biomedicine are given to instructing and "correcting" locals, not so much as a form of education, but rather as part of the paternalistic model of taking care of locals and giving them guidelines that are "for their own good," as was demonstrated in interviews with the biomedical physicians and resonated in other conversations with representatives of the Management of the Reserve. In this way, the same political-economic processes driving the management of forest resources are influencing the availability and use of health care services. When the governmental agency created La Casa, local people in Ayotitlán experienced it as a way of the national government (i.e., outsiders) taking control over their medicinal plants and traditional healing practices. The government was taking up traditional medicine and making it fit into a national model of community development, which the people of Ayotitlán continue to resist.

There is an ongoing debate within the arena of conservation as to how to best achieve conservation and human well-being goals. The big question, in many cases the ultimate goal, is can we improve local well-being and health while decreasing the extent of environmental degradation? Doing so would be a "win-win" approach, and includes a whole cornucopia of strategies, such as: integrated conservation and development projects, community-based conservation, CBPR, debt-for-nature swaps, and so forth (McShane et al. 2011). As we know, however, there are more examples of failed "win-win" strategies than there are success stories. Instead, I suggest that in the case of La Casa, we need to reframe the goals and think about conservation that improves local human well-being and environmental sustainability as "trade-offs" (2011). By being more realistic and acknowledging the hard-choices and trade-offs being made by different stakeholders and parties in development, the goals will be more attainable. I am going to organize my analysis of La Casa around these sets of tensions, which might illustrate how we can shift the rhetoric away from "win-win" towards a trade-offs approach to community-based development programs.

Tension 1: Scale

The first area that tensions emerge at La Casa is looking at the scale: the social, ecological, and economic assessments of so-called development are different if we look at Ayotitlán versus a national perspective—or, as my research showed, even within the same Reserve. The community of Ayotitlán is the most marginalized in the Reserve and is a place where the tension between local people and governmental development programs is palpable. The desire for Ayotitlán to "develop" along a path of privatization and modernization is the dominant model for community development in Mexico but is actively resisted by local people who choose not to participate.

The cookie-cutter approach of the national government programs, and the relative power of biomedicine in Ayotitlán, is demonstrated by an interview I conducted with Rodolfo who was a twenty-three-year-old medical physician completing his social service in Ayotitlán at the time of my research. When I interviewed Rodolfo, he was pessimistic about the use and viability of traditional medicine in Ayotitlán—noting primarily that he thinks that they have lost the knowledge of how to use medicinal plants. He said that traditional culture in Ayotitlán has been completely eroded and the community needs more exposure to modernity in order to learn to appreciate modern technology and want more things. According to Rodolfo's logic, the more the community members want to consume, the more driven they will be to work and buy those things. From his perspective, the governmental program OPORTUNIDADES (a social welfare program) was the primary culprit in creating what he said was a welfare-dependent community. He commented that people in Ayotitlán were "too lazy, simply having more and more babies and waiting for their government stipends each month."

Large scale surveys and program evaluations that occur at the national level tend to obscure local-lived realities. This is seen in the way that the biomedical model of health has permeated the healthcare options available in Ayotitlán—and reflects contradicting national level policies (i.e., those that encourage retention of indigenous cultural heritage, and those that promote the spread of biomedicine).

Tension 2: Context

Next, we need to consider the trade-offs and hard choices of sustainable development-type programs in terms of the historical and social contexts. The local history of failed experiences and interactions with sustainable development projects perhaps plagued the process of establishing and maintaining La Casa. The community members who were involved with La Casa may not have been broadly representative, or sufficiently integrated, with Ayotitlán traditional modes of leadership—such as the Consejo de Mayores (Council of Elders). Recall that the individual who was directly responsible for La Casa had become an outsider to the community through achieving advanced education and working for the national government. Thus, this person was no longer viewed as an indigenous person from a neighboring community, but rather as another extension of the national government which people in Ayotitlán view as trying to take possession of local cultural heritage.

Consuelo, a curandero I interviewed in Ayotitlán, told me that he was making a modest living from curing with medicinal plants. He emphasized that he only charges the bare minimum, often accepting payments in kind. Consuelo complained about traditional healers in nearby communities who charge excessively (because they can; in more removed communities there is no access to emergency medical services and things like scorpion stings and infectious diarrheal or fever diseases are common). Consuelo charged very little because the people don't have much money and "someone has to cure them." But, he also said he uses medicinal plants because they are good for curing and they are better than modern medicine, more effective.

The local decision-making and governance bodies are frequently co-opted by other governing agencies in the Reserve. As I found through my interviews, the biomedical practitioners who are planted in the community from outside bring with them a set of ideologies that can conflict with local cultural values. Contrast the perspectives of the local curandero to the biomedical physician, who said he saw no benefits of medicinal plant use for himself or for others. He is more confident in modern medicine because "it is based on science and is simply better." He feels that "it is good that I am here to provide medicines for people and to help them learn more about modern medicine."

There is a great deal of skepticism of outsiders in Ayotitlán—and a real fear of ongoing theft of their traditional knowledge and environmental resources. The fear and skepticism is emerging from a long and ongoing history of exploitation of natural resources in this area. (Perhaps part of the double-edged sword of becoming and being a biosphere reserve.)

Tension 3: Pluralism

The third tension I want to consider is pluralism. Medical pluralism in Mexico is expansive and reflects a national identity that has been built on the concept of *Mestizaje* (or the *blending of races*). The male nurse, Efren, working full-time at the biomedical clinic in Ayotitlán best embodies the mestizaje and medical pluralism of the community. When I asked Efren if he thinks that many community members use medicinal plants he was baffled by the question. "Of course they use medicinal plants—they are an indigenous community!" The people there still know a lot about plants and that is how they are used to curing themselves and it is their preferred mode of curing. Yes, people in Ayotitlán also come to the clinic, but usually after the person is really sick or only if they don't know what is wrong. But they can usually cure themselves using plants, according to Efren.

Efren was thirty-three years old and originally from a nearby indigenous community. He and his wife had moved to Ayotitlán and were raising their young toddler there. He said that it was good that community members use medicinal plants. The people value medicinal plants, have a great deal of knowledge about plant uses, and have a certain amount of pride in their medicinal plant repertoire. In fact, Efren told me that he has used medicinal plants himself. Efren told me that he personally benefits from the use of medicinal plants because they are "very healthy and effective in promoting health." Efren saw that community members benefited from the use of medicinal plants because of their accessibility and the capitalization on local cultural knowledge. There really are no disadvantages to using medicinal plants, according to Efren. It is part of the local culture and identity.

Individuals in Ayotitlán, like indigenous people throughout Mexico, make conscious decisions to employ both traditional medicine and biomedicine. However, in the case of La Casa, community members were obstructed from making this choice because of very real ramifications. Community members in Ayotitlán are exercising the principle of making a hard choice in the trade-off between using medicinal plants and traditional healing practices and the biomedical services provided at the local clinic. The hard choice that individuals had to make in terms of the therapeutic option was also demonstrated by the daughter of a curandero (healer) in Ayotitlán who told me her reason for going to the regional hospital for the birth of her child was because she *could*, but not (she emphasized) because she did not trust traditional midwives. Instead, she knew that she and her baby would have a difficult transfer into biomedical clinics later if she had chosen not to have her baby in the hospital. She was avoiding a conflict or tension that she perceived between biomedicine and traditional medicine. Like most people in Ayotitlán, she wants to have the best of both worlds, and that was what La Casa was intended to help provide.

Tension 4: Complexity

Lastly, we arrive at the issue of complexity. Anthropogenic relationships with environments are complex and intricate. The real essence of the complexity in the case of La Casa, however, is really about the goals of supporting local cultural heritage, environmental stewardship, and promoting health in a community that is under pressure to "develop" along a Western model of modernization. There are many tensions that emerge from trying to achieve so many goals, of course, and trade-offs are being made.

In the context of institutionally supported community development projects, there is a push to administer static programs and management strategies. As conservation and development scholars frequently point out, the relationships between humans and environments are constantly in flux and in processes of change, which makes adaptive management strategies more appealing.

La Casa might have been more successful if there had been flexibility in the scope of services that it could provide; the sources of financial security had not been linked to one governmental agency (and funneled through one individual researcher); and if the relationships with other health care providers who work in the community had been better negotiated and maintained throughout the process. Perhaps a benefit that can come from the experience of La Casa de Medicina Tradicional will be to highlight some of these hard choices and trade-offs.

As illustrated in figure 9.1, the environmental goals were to encourage the retention and utility of indigenous knowledge. However, one question about this strategy is whether or not ex situ gardens will have the same conservational value as in situ or wild plant resources. Understanding the forest as a whole, and the ecosystem relationships in it, is critical to sustainable resource management. Recall Consuelo, the curandero in Ayotitlán, who reported to me that he attends educational seminars across Mexico (Puebla, Veracruz, and Guerrero). According to Consuelo, "in those regions there is actually more knowledge of plants. The plants are dying and disappearing in Ayotitlán."

Benefit Categories	**Environmental conservation**	**Human well-being**
Gain	Supports retention and use of traditional medicinal plant knowledge as part of a rubric of natural resource knowledge that can be used to support sustainable resource uses.	Demonstrates support and value of traditional knowledge and cultural heritage. Opportunities for holistic health services that emphasize and value cultural traditions and less-intensive interventions.
Loss	New construction of house and gardens decreases value of *in situ* and wild plant resources.	Feel pressured to choose between using biomedical clinic or traditional medicine, but not both.

Figure 9.1. Reality of trade-offs in example of La Casa.

Summary of Key Findings

The story of La Casa de Medicina Tradicional embodies the important lessons that are found in this ethnography. I embarked to study the relationship between economic livelihoods, individuals' health experiences, and traditional medicinal plant knowledge. This ethnography documents a detailed exploration of these themes in the Reserve. The threads of indigenous people, conservation and development, and traditional ecological knowledge are interwoven in the lives and livelihoods of people living in the Reserve.

Conducting research in the Reserve required a great deal of preparation and preliminary fieldwork. As I immersed myself in the local cultures and engaged with the study communities as a visitor and (later) anthropologist, I learned about the rich cultural, social, and political history of the area. The types of interactions within the communities themselves, and between communities and outside agents, are related to the type of health decisions that individuals make.

Shared systems of land tenure (such as ejidos and comunidades indígenas) along with their traditional governance structures (like the Consejo de Mayores) contribute to the local indigenous identity. As I discussed in chapter 5, for some outsiders, this can be viewed as a barrier to so-called development.

At the outset of my study, I projected that communities where the primary source of subsistence for households was based in farming activities, those communities would also boast longer lists of medicinal plants on average. My findings did seem to corroborate this hypothesis in some ways, since the community with the highest reliance on cattle ranching (Zacualpan) also had the shortest length lists of medicinal plants on average. However, this is only one portion of the story about medicinal plant knowledge and medical pluralism that I have unfolded.

In exploring the variety of health and healing options that members of the three study communities used, I found that some individuals valued medicinal plants very highly but for varying reasons also used modern medicine. The result is a medically pluralistic society in which individuals and families are incorporating both traditional and modern medicines into their repertoire of treatment options. The medical pluralism resonates with the mestizaje, or blending of Spanish and indigenous culture, which is a dominant aspect of the local identity.

The fast pace of globalization and sweeping modernization or development programs have had profound impacts in the Reserve. Throughout each aspect of this ethnographic study, we are reminded of the presence of outsiders and development programming. The auspices of these outsiders and the disparate programs related to conservation and natural resource management have borne many different names and hats, but the processes and outcomes bear many similarities. The lessons that we can take from this ethnography are demonstrated in the example of La Casa de Medicina Tradicional in Ayotitlán.

Anthropology has dealt with issues of health and conservation from many angles. At the end of the day, we still have not fully answered some critical questions, and within academia there is often a tendency to mirror the divisions that exist in sustainable development planning: health and environmental issues are often treated as separate spheres. There are many opportunities for increasing inter- and trans-disciplinary approaches that emphasize collaboration in problem-solving.

Glossary

Alloparenting

System of child-rearing in which adults other than the parents (usually family members) aid in the primary activities of caring for young children.

Animal Husbandry

Raising domesticated animals for the use of secondary products such as milk, eggs, meat, labor, or blood.

Calpulli

Nahuatl. A system of community organization and governance that existed principally in the Nahua cultures in central Mexico. It was a main historical and cultural reference point for the reestablished communal land tenure system of *ejidos.*

Community Capacity Building

Refers to the goals of various types of community development projects in which equity in access to social and economic resources is increased. The underlying principle is to help communities to help themselves through improved infrastructure (such as schools, health clinics, roads, etc.) and resources (vocational training programs, job security, food and nutritional supplements, etc.).

Curandero

Spanish. A general term used to refer to a traditional healer in Mexico and parts of Latin America. The term comes from the verb *curar* which means "to heal" in Spanish.

Ejidatario

Spanish. A voting member of an *ejido* with all of the rights and responsibilities of participation in decision-making and shared ownership of the *ejido*.

Ejido

Spanish. A land tenure system established in 1934 in Mexico with the primary intention of redistributing lands that had been seized in the formation of the encomienda system and haciendas during previous centuries. The structure of the ejido is intended to resemble the ancient Nahau system of collectively owned and managed resources. Until 1994, ejido land could not be sold or transferred to non-community members.

Emic

Term used to describe the internal and culturally-based nature of a concept; an authentic or insider's understanding.

Encomienda

Spanish. A system of land and human management established by the Spanish during the colonization of current-day Mexico and other areas in Latin America. The Spanish crown administered the encomienda system which was organized similarly to the European feudal system: land was managed by a designee of the Royal Crown, native peoples and peasantry were assigned to work on the encomienda and in exchange their basic daily needs were intended to be met by the overlord. The encomienda system was abolished in the 1917 Constitution of Mexico after the Revolution.

Ethnomedical System

General term for an organized and ordered body of knowledge, practices, and relationships between individuals that is focused on the maintenance and restoration of health.

Ex Situ

Refers to the process of studying plants or other natural resources away from the habitat in which they are found.

Folk Medicine

An organized system of health maintenance that has a loose professional hierarchy and is highly integrated with the dominant cultural and social system in which it is embedded. Also called traditional medicine.

Freelist

Quantitative research technique that is a structured interview and is frequently used by ethnobiologists to identify the contents of a cognitive domain. For example, an informant is asked to list all the types of shoes that he or she can think of. The goal is to be exhaustive in identifying the entire cognitive domain.

Globalization

Rapid and multifaceted diffusion of cultural mores, information and ideas, and natural and economic resources. Various definitions of globalization exist owing to the complexity of the concept and the scope of factors that it implies.

Hacienda

Spanish. A large plantation or factory that operates as a company for the production of agricultural products, raw natural resources, or manufactured goods. The hacienda system is based on economic productivity and was established in Mexico and Latin America during the period of colonization by Europeans.

In Situ

Refers to the process of studying plants or other natural resources in the habitat in which they are found.

Market Participation

Economic and social activities related to purchasing and sale of material goods at all levels: community, regional, national, and global.

Medicinal Plant Knowledge

A culturally defined category of the utility of a natural resource for the purposes of health maintenance.

Milpa

Spanish. Traditional system of crop interplanting common throughout Mexico and Latin America. Based on maize, squash, and chile or beans, depending on the specific location where it is being practiced. These agricultural products also form the basis of local diet and are usually planted for autoconsumption within the household and not for market sale.

NAFTA

The North American Free Trade Agreement which was put into effect in 1994 among the United States, Canada, and Mexico. Though there have been multiple impacts of the treaty, one of the primary impacts was to open Mexican markets for the consumption of goods from the United States and the cheap sale of natural and human resources to the United States.

Naturistas

Spanish. A type of health practitioner who usually does not have an advanced degree in allopathic medicine, but who was trained through apprenticeship and extensive hands-on study to use plant-based remedies, traditional therapies, and/or other types of non-biomedical approaches to restoring and maintaining health.

Parcela

Spanish. Plot of land or field that is used for any type of agricultural production.

Protected Area

Geographically designated location where human activities are restricted, most often for the purpose of preserving and supporting biodiversity. This status is politically determined and implies a structured governance system, which varies according to the many sub-types of protected areas and the explicit purpose of each unique protected area.

Traditional Agroecosystem

Productive area that is managed according to indigenous and cultural knowledge and methods for the purposes of satisfying human crop and animal needs while maintaining an integrative and inclusive approach.

Popular Medicine

Term used to refer to the common health and medicine practices that are held by most members of the cultural group and which serve as the basis for resolving the majority of everyday ailments or health complaints.

Professional Medicine

A highly organized and elaborate system of health and medical knowledge that has a unique cultural, social, and political system of its own. The cultural and social organization of professional medicine includes specific roles which are often hierarchical and all of the elements of a cultural system, including rites of passage and specialized knowledge.

Bibliography

Agrawal, Arun, and Clark C. Gibson. 1999. "Enchantment and disenchantment: The role of community in natural resource conservation." World development 27(4):629-649.

Ahmed, Selena, John R. Stepp, Robban A. J. Toleno, and Charles M. Peters. 2010. "Increased Market Integration, Value and Ecological Knowledge of Tea Agroforests in the Akha Highlands of Southwest China." Ecology and Society 15(4):27.

Alcorn, Janis B. 2001. "Good Governance, Indigenous Peoples and Biodiversity Conservation: Recommendations for Enhancing Results Across Sectors." In Biodiversity Support Program: WWF.

Anand, Sudhir, and Amartya Sen. 2000. "Human development and economic sustainability." World Development 28(12):2029-2049.

Anderson, Eugene N. 2002. "Biodiversity Conservation: A New View from Mexico." In Seventh International Congress of Ethnobiology, edited by John R. Stepp, Felice S. Wyndham and Rebecca K. Zarger. Athens, GA: The International Society of Ethnobiology.

Appadurai, Arjun. 1996. Modernity at large: The cultural dimensions of globalization. Minnepolis: University of Minnesota Press.

Argyrou, Vassos. 2005. The logic of environmentalism: Anthropology, ecology, and postcoloniality. New York: Berghahn.

Aron, Raymond. 1998. Main Currents in Sociological Thought: Durkheim, Pareto, Weber. Vol. 2: Transaction Publishers.

Ayantunde, Augustine, Mirjam Briejer, Pierre Hiernaux, Henk Udo, and Ramadjita Tabo. 2008. "Botanical Knowledge and its Differentiation by Age, Gender and Ethnicity in Southwestern Niger." Human Ecology 36(6):881-889.

Benz, Bruce F. 1986. Taxonomy and evolution of Mexican maize. Ph.D. dissertation. University of Wisconsin.

Benz, Bruce F., Francisco Santana M., Rosario Pineda L., Judith Cevallos E., Luis Robles H., and Domitila de Niz L. 1994. "Characterization of mestizo plant use in the Sierra de Manantlán, Jalisco-Colima, Mexico." Journal of Ethnobiology 14(1):23-41.

Benz, Bruce F., Hugo Perales, and Stephen Brush. 2007. "Tzeltal and Tzotzil Farmer Knowledge and Maize Diversity in Chiapas, Mexico." Current Anthropology 48(2).

Benz, Bruce F., Judith Cevallos, Francisco Santana, Jesus Rosales, and Sergio Graf. 2000. "Losing knowledge about plant use in the Sierra de Manantlán Biosphere reserve, Mexico." Economic Botany 54:183-191.

Berger, Peter L., and Thomas Luckmann. 1966. The Social Construction of Reality: A Treatise in the Sociology of Knowledge. Garden City, NY: Anchor Books.

Berger, Sebastian. 2008. "K. William Kapp's theory of social costs and environmental policy: Towards political ecological economics." Ecological Economics:244-252.

Berkes, Fikret. 1999. Sacred Ecology: Traditional Ecological Knowledge and Resource Management. New York: Taylor & Francis.

Berkes, Fikret, Carl Folke, and Madhav Gadgil. 1995. "Traditional Ecological Knoweldge, Biodiversity, Resilience and Sustainability." In Biodiversity Conservation: Problems and Policies, edited by Charles A. Perrings, Karl-Göran Mäler, Carl Folke, C. S. Holling and Beng-Owe Jansson. Dordrecht Kluwer Academic Publishers.

Berlin, Brent. 1992. Ethnobiological classification: principles of categorization of plants and animals in traditional societies. Princeton, N.J.: Princeton University Press.

Berlin, Brent, and Elois Ann Berlin. 2004. "Community Autonomy and the Maya ICBG Project in Chiapas, Mexico: How a Bioprospecting Project that Should Have Succeeded Failed." Human Organization 63(4):472-486.

Bernard, H. Russell. 1995. Research Methods in Anthropology: Qualitative and Quantitative Approaches. 2nd edition. New York: Altamira Press.

Björk, Peter. 1997. "Marketing of finnish eco-resorts." Journal of Vacation Marketing 3(4):303-313.

Björk, Peter. 2000. "Ecotourism from a conceptual perspective, and extended definition of a unique tourism form." International Journal of Tourism Research 2(3):189-202.

Boas, Franz, and George W. Stocking. 1989. A Franz Boas Reader: The Shaping of American Anthropology, 1883-1911. Chicago: University of Chicago Press.

Boo, Elizabeth. 1990. Ecotourism: The Potentials and Pitfalls. Washington, D.C.: World Wildlife Fund.

Borgatti, Stephen P. 1996. ANTHROPAC 4.0 Methods Guide. Natick, MA: Analytic Technologies.

_____. 1998. ANTHROPAC 4.983/X. Natick, MA: Analytic Technologies.

Boster, James. 1986. Exchange of varieties and information between Aguaruna manioc cultivators. American Anthropologist 88:429-36.

Bramwell, Bill, and Bernard Lane. 1993. "Interpretation and sustainable tourism: the potential and the pitfalls." Journal of Sustainable Tourism 1(2):71-80.

Brechin, Steven R., Peter R. Wilshusen, Crystal L. Fortwangler, and Patrick C. West. 2002. "Beyond the Square Wheel: Toward a More Comprehensive Understanding of Biodiversity Conservation as Social and Political Processes." Society and Natural Resources 15:41-64.

Brockmann, C., Jacqueline Gónzalez O., and Oscar Gónzalez O. 1994. Cuzalapa y Ayotitlán: el derecho de los vencidos, quinientos años despúes (Informe del servico social prestado en la reserva de la biosfera Sierra de Manantlán). Guadalajara: ITESO.

Brown, Michael, and Barbara Wyckoff-Baird. 1995. Designing Integrated Conservation and Development Projects. Washington, DC: Biodiversity Support Program.

Buenz, Eric J. 2005. "Country development does not presuppose the loss of forest resources for traditional medicine use." Journal of Ethnopharmacology 100(1-2):118-23.

Butler, Richard W. 1990. "Alternative tourism: pious hope of Trojan horse?" Journal of Travel Research 28(3):40-45.

Butler, Richard W., and Tom D. Hinch. 1996. Tourism and Indigenous Peoples Boston: International Thomson Business Press.

Byron, Elizabeth M. 2003. Market integration and health: the impact of markets and acculturation on the self-perceived morbidity, diet, and nutritional status of the Tsimane'Amerindians of lowland Bolivia. Ph.D. dissertation, University of Florida.

Canclini, Nestor Garcia. 1995. Hybrid Cultures: Strategies for Entering and Leaving Modernity. Minneapolis-London: University of Minnesota Press.

Caniago, Izefri, and Stephen F. Siebert. 1998. "Medicinal plant ecology, knowledge and conservation in Kalimantan, Indonesia." Economic Botany 52:229-50.

Carabias Lillo, Julia, Enrique Provencio, Javier de la Maza Elvira, and Sergio Graf. 2000. Programa de Manejo Reserva de la Biosfera Sierra de Manantlán. Mexico City: Instituto Nacional de Ecologia.

Ceballos-Lascuráin, Héctor. 1987. "The future of 'ecotourism'." Mexico Journal (January):13-14.

_____. 1998. "Introduction." In Ecotourism: A guide for planners and managers, Volume 2, edited by Kreg Lindberg, Megan Epler Wood, and David Engeldrum. North Bennington, VT: The Ecotourism Society.

Chambers, Robert. 1997. Whose Reality Counts? Putting the First Last. London: Intermediate Technology Publications.

Chant, Sylvia. 1991. Women and survival in Mexican cities: Perspectives on gender, labour, markets and low-income households. New York: Manchester University Press.

CIDOC. 2005. Current Housing Situation in Mexico 2005. Centro de Investigacion y Documentacion de la Casa (CIDOC), Mexico City.

Collier, George A, and Elizabeth Lowery Quaratiello. 2005. Basta! Land and the Zapatista Rebellion in Chiapas. 3rd edition. Oakland, California: Food First Books.

Cox, Paul A. 2000. "Will tribal knowledge survive the millennium?" Science 287(5450):44-45.

Cushing, Frank Hamilton, and Jesse Green. 1979. Zuñi: selected writings of Frank Hamilton Cushing. Lincoln: University of Nebraska Press.

Davis, Benjamin, Guy Stecklov, and Paul Winters. 2002. "Domestic and international migration from rural Mexico: Disaggregating the effects of network structure and composition." Population Studies 56(3):291-309.

De Walt, Billie R. 1994. "Using indigenous knowledge to improve agriculture and natural resource management." Human Organization 53(2):123-131.

Delegados de Los Pueblos Indígenas. 2003. Declaracion de la Sierra de Manantlán. Ayotitlán, Territorial Náhuatl, Jalisco, Mex.:Congreso Nacional Indígena (CNI).

Devlin, John F., and Nonita T. Yap. 2008. "Contentious politics in environmental assessment: blocked projects and winning coalitions." Impact Assessment and Project Appraisal 26(1):17-27.

Dion, Michelle. 2005. "The Political Origins of Social Security in Mexico during the Cardenas and Avila Camacho Administrations." Mexican Studies 21(1):59-95.

Escobar, Arturo. 1995. Encountering Development. Princeton, N.J.: Princeton University Press.

Fennell, David A. 2008. Ecotourism. London: Routledge.

Ferguson, James. 1994. The anti-politics machine: "development," depoliticization, and bureaucratic power in Lesotho. Minneapolis: University of Minnesota.

Finkler, Kaja. 1994a. Spiritualist Healers in Mexico: Successes and Failures of Alternative Therapeutics. Salem, Wisconsin: Sheffield Publishing Company.

_____. 1994b. Women in Pain: Gender and Morbidity in Mexico. Philadelphia: University of Pennsylvania Press.

_____. 2001. Physicians at Work, Patients in Pain: Biomedical Practice and Patient Response in Mexico.2nd Edition. Durham, N.C.: Carolina Academic Press.

Fletcher, Robert. 2008. "Living on the Edge: The Appeal of Risk Sports for the Professional Middle Class." Sociology of Sport Journal 25(3):310-330.

_____. 2009. "Ecotourism discourse: challenging the stakeholders theory." Journal of Ecotourism 8(3):269-285.

_____. 2010. "Neoliberal Environmentality: Towards a Poststructuralist Political Ecology of the Conservation Debate." Conservation and Society 8(3):171-181.

Ford, Richard I. 1985. "Anthropological Perspective of Ethnobotany in the Greater Southwest." Economic Botany 39(4):400-415.

Garcia, Maribel Osorio, and Marcelino Castillo Nechar. 2006. Ensayos teorico-metodologicos del turismo: cuatro enfoques. Toluca, Mexico: Universidad Autonoma del Estado de Mexico.

Garcia-Quijano, Carlos G. 2007. "Fishers' Knowledge of Marine Species Assemblages: Bridging between Scientific and Local Ecological Knowledge in Southeastern Puerto Rico." American Anthropologist 109:529–536.

Garro, Linda C. 1986. "Intracultural variation in folk medicinal knowledge: A comparison between groups." American Anthropologist 88:351-370.

_____. 1998a. "On the rationality of decision-making studies: Part 1: Decision models of treatment choice." Medical Anthropology Quarterly 12(3):319-340.

_____. 1998b. "On the rationality of decision-making studies: Part 2: Divergent rationalities." Medical Anthropology Quarterly 12(3):341-355.

Geertz, Clifford. 2000. Available Light: Anthropological Reflections on Philosophical Topics. Princeton: Princeton University Press.

Gerritsen, Peter R. W. 1998. "Community development, natural resource management and biodiversity conservation in the Sierra de Manantlán biosphere, Mexico." Community Development Journal 33(4):314-324.

_____. 2002. Diversity at stake. A farmers' perspective on biodiversity and conservation in Western Mexico. Wageningen Studies on Heterogeneity and Relocalisation 4.

_____. 2004. Estilos agrarios y la forestería comunitaria. Autlán: Universidad de Guadalajara, Centro Universitario de la Costa Sur.

_____. 2010. Perspectivas campesinas en el manejo de los recursos naturals. Mexico City: Mundi-Prensa.

Godoy, Ricardo. 2001. Indians, markets, and rainforests: theory, methods, analysis. New York: Columbia University Press.

Godoy, Ricardo, Nicholas Brokaw, David Wilkie, Daniel Colon, Adam Palermo, Suzanne Lye, and Stanley Wei. 1998. "On trade and cognition: Markets and the loss of folk knowledge among the Tawahka Indians." Journal of Anthropological Research 54:219-33.

Godoy, Ricardo, Victoria Reyes-García, Elizabeth Byron, William R. Leonard, and Vincent Vadez. 2005. "The Effect of Market Economies on the Well-Being of Indigenous Peoples and on Their Use of Renewable Natural Resources." Annual Review of Anthropology 34:121-138.

Gonzalez, Alma A., and Ronald Nigh. 2005. "Smallholder participation and certification of organic farm products in Mexico." Journal of Rural Studies 21(4):449-460.

González-Rossetti, Alejandra, and Olivia Mogollon. 2000. Enhancing the Political Feasibility of Health Reform: The Mexico Case, School of Public Health, Harvard, Cambridge.

Good, Charles M. 1987. Ethnomedical systems in Africa: patterns of traditional medicine in rural and urban Kenya. New York: Guilford Press.

Graeber, David. 2011. Debt: the first 5,000 years. Brooklyn, N.Y.: Melville House.

Graf, Sergio, Eduardo Santana C., Enrique Jardel, Martin Gomez, and Salvador Garcia Ruvalcaba. 2003. "La Reserva de la Biosphere Sierra de Manantlán, Mexico." In Capacidades necesarias para el manejo de areas protegidas: America Latina y el Caribe, edited by Julia Carabias, Javier De la Maza and Rosaura Cadena, 135-153. Arlington, VA: The Nature Conservancy.

Grant, Robert M. 1996. "Prospering in Dynamically-Competitive Environments: Organizational Capability as Knowledge Integration." Organization Science 7(4):375-387.

Guest, Greg. 2002. "Market integration and the distribution of ecological knowledge within an Ecuadorian fishing community." Journal of Ecological Anthropology 6:38-49.

Gutierrez, Marisa, and Peter R. W. Gerritsen. 2009. Turismo rural sustentable en la Costa Sur de Jalisco. Guadalajara: Universidad de Guadalajara, Centro Universitario de la Costa Sur.

Haenn, Nora. 2002. "Nature Regimes in Southern Mexico: A History of Power and Environment." Ethnology 41(1):1-26.

Halffter, Gonzalo. 1994. "Conservación de la biodiversidad y áreas protegidas en los países tropicales." Ciencias 036.

_____. 1995. "Reserva de la biósfera y conservación de la biodiversidad en el siglo XXI." Ciencias 039.

Hall, Colin M., and Stephen W. Boyd. 2003. Ecotourism in Peripheral Areas. Clevedon, UK: Channel and View Publications.

Hamilton, Sarah. 2002. "Neoliberalism, Gender, and Property Rights in Rural Mexico." Latin American Research Review 37(1):119-143.

Harrison, Margaret E. 2001. "Mexican Physicians, Nurses and Social Workers." In Mesoamerican Healers, edited by Brad R. Huber and Alan R. Sandstrom, 270-306.

Harshbarger, John. 1896. "The Purposes of Ethno-botany." Botanical Gazette 21(3):146-154.

Higham, James E. S. 2007. Critical issues in ecotourism understanding a complex tourism phenomenon. Oxford: Elsevier.

Higham, James E. S., and Anna Carr. 2003. "Defining ecotourism in New Zealand: differentiating between the defining parameters within a national/regional context." Journal of Ecotourism 2(1):17-32.

Hoffman, David M. 2009. "Institutional Legitimacy and Co-Management of a Marine Protected Area: Implementation Lessons from the Case of Xcalak Reefs National Park, Mexico." Human Organization 68(1):39-54.

Honey, Martha. 2003. "Summary of major principles regarding tourism and indigenous peoples and local communities." In Rights and Responsibilities: A Compilation of Codes of Conduct for Tourism and Indigenous and Local Communities, edited by M. Honey and S. Thullen. Center on Ecotourism and Sustainable Development and the International Ecotourism Society.

_____. 2008. Ecotourism and sustainable development: Who owns paradise? 2nd Edition. Washington, D.C.: Island Press.

Icamina, Paul. 1993. "Threads of common knowledge." International Development Research Centre Report 21(1):14-6.

Igoe, James, and Daniel Brockington. 2007. "Neoliberal conservation: A brief introduction." Conservation and Society 5(4): 432-449.

IMECBIO. 2000. Programa de Desarrollo Regional Sustentable: Región de la Sierra de Manantlán, Estados de Jalisco y Colima. In Instituto Manantlán de Ecología y Conservación de la Biodiversidad, edited by Enrique Jardel. Autlán de Navarro: Centro Universitario de la Costa Sur, Universidad de Guadalajara

IMSS. Instituto Mexicano del Seguro Social. IMSS 2005. Cited 31 January 2005. Available from http://www.imss.gob.mx/IMSS/

INE. 2000. Áreas Naturales Protegidas de México. México City; SEMARNAP, INE (Instituto Nacional de Ecología).

INEGI. 2005. ITER2005. Sistemas Nacionales Estadístico y de Información Geográfica.

Inglehart, Ronald. 1990. Culture shift in advanced industrial society: Princeton University Press.

Israel, Barbara A., Amy J. Schulz, Edith A. Parker, and Adam B. Becker. 1998. "Review of Community-Based Research: Assessing Partnership Approaches to Improve Public Health." Annual Review of Public Health 19:173-202.

Jantzi, Terrence Lee. 2000. Local program theories and social capital: a case study of a non-governmental organization in eastern Bolivia. http://worldcat.org. Accessed on January 25, 2012.

Jardel, Enrique. 1992. Estrategia para la conservacion de la Reserve de la Biosfera Sierra de Manantlan. Editorial. Universidad de Guadalajara.

Johnston, A.M. 2006. Is the Sacred for Sale? Tourism and Indigenous Peoples. London: Earthscan.

Kleinman, Arthur. 1978. "Concepts and a model for the comparison of medical systems as cultural systems." Social Science and Medicine 12(2B):85-95.

_____. 1980a. "Chapter 6: Family-Based Popular Health Care." In Patients and Healers in the Context of Culture, edited by Arthur Kleinman. Berkeley: University of California Press.

_____. 1980b. Patients and Healers in the Context of Culture: An Exploration of the Borderland between Anthropology, Medicine, and Psychiatry. Berkeley: University of California Press.

Krauze, Enrique. 1998. Mexico: Biography of Power: A History of Modern Mexico, 1810-1996. Translated by Hank Heifetz. 1st HarperPerennial edition New York: HarperPerennial.

Laitner, Karen, and Bruce P. Benz. 1994. "Las condiciones culturales y ambientales en la reserva de la biosfera Sierra de Manantlán en tiempo de la conquista: una perspective de los documentos etnohistóricos secundarios." Estudios del hombre 1:15-45.

Lassey, Marie L., William R. Lassey, and Martin J. Jinks. 1997. Health Care Systems Around the World: Characteristics, Issues, Reforms. Upper Saddle River, NJ: Prentice Hall.

Leeper, James D. 2008. "What statistical analysis should I use? Statistical analyses using SPSS." UCLA: Academic Technology Services, Statistical Consulting Group.

Li, Tania. 2007. The will to improve: governmentality, development, and the practice of politics. Durham: Duke University Press.

Long, Norman, and Ann Long. 1992. Battlefields of knowledge. The interlocking of theory and practice in social research and development. London and New York: Routledge Publishers.

Lu, Flora. 2007. "Integration into the market among indigenous peoples: A cross-cultural perspective from the Ecuadorian Amazon." Current Anthropology 48(4):593-602.

Mann, Charles C. 2005. 1491: New Revelations of the Americas Before Columbus. New York: Knopf.

Martinez R., Luis M., Peter R.W. Gerritsen, Ramon Cuevas, and Jesus Rosales A. 2006. "Incorporating principles of sustainable development in research and education in western Mexico." Journal of Cleaner Production 14:1003-1009.

McCauley, D.J. 2006. "Selling out on nature." Nature 443(7):27-28.

McGehee, Nancy Gard, and Alison C. Meares. 1998. "A Case Study of Three Tourism-related Craft Marketing Cooperatives in Appalachia: Contributions to Community." Journal of Sustainable Tourism 6(1):4-25.

McNeeley, Jeffrey A., Kenton R. Miller, Walter V. Reid, Russell A. Mittermeier, and Timothy B. Werner. 1990. Conserving the world's biological diversity. IUCN, WRI, CI, WWF-US, The World Bank: Washington, D.C.

McShane, Thomas O., Paul D. Hirsch, Tran Chi Trung, Alexander N. Songorwa, Ann Kinzig, Bruno Monteferri, David Mutekanga, Hoang Van Thang, Juan Luis Dammert, Manuel Pulgar-Vidal, Meredith Welch-Devine, J. Peter Brosius, Peter Coppolillo, and Sheila O'Connor. 2011. "Hard choices: Making trade-offs between biodiversity conservation and human well-being." Biological Conservation 144(3):966-972.

Merrill, Tim L., and Ramón Miró. 1996. Mexico: a country study, Federal Research Division. Washington D.C.: The Library of Congress.

Minkler, Meredith. 1997. Community Organizing and Health. N.J.: Rutgers University Press.

_____. 2004. "Ethical Challenges for the "Outside" Researcher in Community-Based Participatory Research." Health Education & Behavior 31(6):684-697.

Moore, Jerry D. 2008. Visions of culture: An introduction to anthropological theories and theorists. Lanham, M.D.: Rowman & Littlefield.

Morgan, Lynn. 1993. Community Participation in Health: The Politics of Primary Care in Costa Rica. Cambridge: Cambridge University Press.

Mowforth, Martin, and Ian Munt. 2003. Tourism and Sustainability: development and new tourism in the Third World. New York: Routledge.

Napolitano, Valentina. 2002. Migration, Mujercitas, and Medicine Men: Living in Urban Mexico. Berkeley: University of California Press.

National Center for Health Statistics. 1998. Health in the Americas. Edited by CDC.

Nolan, Justin M., and Michael C. Robbins. 2000. "Cultural conservation of medicinal plant use in the Ozarks." Human Organization 58(1):67-72.

Oldfield, Margery L., and Janis B. Alcorn. 1987. "Conservation of traditional agroecosystems." BioScience 37(3):199-208.

Olson, Elizabeth Anne. 2009. Nahua people of the Sierra of Manantlán Biosphere Reserve: Livelihoods, health experiences, and medicinal plant knowledge in Mexico. Ph.D. dissertation. Case Western Reserve University, Ohio.

_____. 2011. "Notions of rationality and value production in ecotourism: examples from a Mexican biosphere reserve." Journal of Sustainable Tourism 20 (2):215-233.

_____. 2013. "Anthropology and Traditional Ecological Knowledge: A summary of quantitative approaches to traditional knowledge, market participation, and conservation." Culture, Agriculture, Food and Environment 35(2):140-151.

Olson, Elizabeth Anne, and Peter R.W. Gerritsen. 2011. "Some Reflections on Community Development and Outsider Intervention in the Context of Protected Natural Areas in Western Mexico." Ecological and Environmental Anthropology 6(1):10-25.

Orams, Mark B. 1995. "Towards a more desirable form of ecotourism." Tourism Management 16(1):3-8.

Ostrom, Elinor. 1990. Governing the Commons: The Evolution of Institutions for Collective Action. Cambridge: Cambridge University Press.

Page, Stephen J. and Ross K. Dowling. 2002a. "Community-based ecotourism: Management and development issues." In Ecotourism, 244-247. Harlow, UK: Pearson Education.

Page, Stephen J. and Ross K. Dowling. 2002b. Ecotourism. Harlow, UK: Pearson Education.

PAHO. 2003. Mexico: Core Health Data Selected Indicators, Data Updated to 2002. Washington D.C.: Pan American Health Organization, Technical Health Information System.

_____. 2004. Regional Core Health Data Initiative; Technical Health Information System. Washington D.C.: PAHO Health Analysis and Information Systems Area.

Pfeiffer, James, and Mark Nichter. 2008. "What Can Critical Medical Anthropology Contribute to Global Health? A Health Systems Perspective." Medical Anthropology Quarterly 22(4):410-415.

Prance, Ghillean T. 2000. "Ethnobotany and the future of conservation." Biologist (London) 47(2):65-68.

Quinlan, Marsha. 2005. "Considerations for Collecting Freelists in the Field: Examples from Ethnobotany." Field Methods 17(3):219-234.

Redford, Kent H. 1991. "The Ecologically Noble Savage." Orion 9:24-29.

Reyes-Garcia, Victoria. 2001. Indigenous people, ethnobotanical knowledge, and market economy. A case study of the Tsimane' Amerindians in lowland Bolivia. Ph.D. dissertation, University of Florida.

Reyes-Garcia, Victoria, Ricardo Godoy, Vincent Vadez, Lillian Apaza, Elizabeth Byron, Tomas Huanca, William R. Leonard, Edith Perez, and David Wilkie. 2003. "Ethnobotanical knowledge shared widely among Tsimane' Amerindians, Bolivia." Science 299(5613):1707.

Reyes-Garcia, Victoria, Vincent Vadez, Susan Tanner, Thomas McDade, Tomas Huanca, and William R. Leonard. 2006. "Evaluating indices of traditional ecological knowledge: a methodological contribution." Journal of Ethnobiology and Ethnomedicine 2:21.

Reyes-Garcia, Victoria, Vincent Vadez, Tomas Huanca, William R. Leonard, and Thomas McDade. 2007. "Economic development and local ecological knowledge: A deadlock? Quantitative research from a native Amazonian society." Human Ecology 35(3):371-377.

Rigg, Jonathan. 2006. "Land, farming, livelihoods, and poverty: Rethinking the links in the Rural South." World Development 34(1):180-202.

Rojas, Rosa, I. Alcocer, H. Fajardo, G. Martínez, M.A. Márquez, E. Miramontes, D. Ortega, C. Palomar, J.C. Pérez, and A. Sandoval. 1996. "La comunidad y sus recursos: Ayotitlán ¿desarrollo sustentable? Caracterización integral del ejido Ayotitlán, Municipio de Cuautitlán, Jalisco." University of Guadalajara/Instituto Nacional Indigenista: 277.

Romney, A. Kimball, Susan C. Weller, and William H. Batchelder. 1986. "Culture as Consensus: A Theory of Culture and Informant Accuracy." American Anthropologist 88(2):313-338.

Rosenthal, Joshua P. 2006. "Politics, Culture, and Governance in the Development of Prior Informed Consent in Indigenous Communities." Current Anthropology 47(1):119-141.

Sachs, Wolfgang. 1992. "Introduction." In The Development Dictionary: A Guide to Knowledge as Power, edited by Wolfgang Sachs, 1-5. London: Zed Books.

Salazar, Noel B. 2012. "Community-based cultural tourism: issues, threats and opportunities." Journal of Sustainable Tourism 20(1):9-22.

Santana Michel, Francisco Javier, and Ramón Cuevas Guzmán. 2007. Personal Communication. Autlán de Navarro, 26 September 2007.

Sheridan, Thomas E. 1988. Where the Dove Calls: The political ecology of a peasant corporate community in Northwestern Mexico. Tucson: The University of Arizona Press.

Sillitoe, Paul. 2007. Local science vs. global science: Approaches to indigenous knowledge in international development. Volume 4, Studies in environmental anthropology and ethnobiology. New York: Berghahn Books.

Sociedad Hipotecaria Federal, edited by Sara Topelson Grinberg. Cambridge, MA: Joint Center for Housing Studies.

Srithi, Kamonnate, Henrik Balslev, Prasit Wangpakapattanawong, Prachaya Srisanga, and Chusie Trisonthi. 2009. "Medicinal plant knowledge and its erosion among the Mien (Yao) in northern Thailand." Journal of Ethnopharmacology 123(2):335-342.

Staff. 1995. "Farmers Devalued with the Peso." The NAFTA and Inter-American Trade Monitor 2(4).

Stepp, John R. 2005. "Advances in Ethnobiological Field Methods." Field Methods 17(3):211-218.

Stepp, John R., Hector Castaneda, and Sarah Cervone. 2005. "Mountains and Biocultural Diversity." Mountain Research and Development 25(3):223–227.

Steward, Julian Haynes. 1955. Theory of culture change; the methodology of multilinear evolution. Urbana: University of Illinois Press.

Stronza, Amanda. 2005. "Hosts and hosts: the anthropology of community-based ecotourism in the Peruvian rainforest." National Association for the Practice of Anthropology Bulletin 23:170-190.

Stronza, Amanda. 2007. "The Economic Promise of Ecotourism for Conservation." Journal of Ecotourism 6(3):210 - 230.

Thompson, Eric C., and Zhang Juan. 2006. "Comparative cultural salience: Measures using free-list data." Field Methods 18(4):398-412.

Thompson, Gary D., and Paul N. Wilson. 1994. "Ejido Reforms in Mexico: Conceptual Issues and Potential Outcomes." Land Economics 70(4):448-465.

TIES. 2004. Definition and ecotourism principles. Washington, D.C.: The International Ecotourism Society.

Toledo, Victor M. 2002. "Ethnoecology: A Conceptual Framework for the Study of Indigenous Knowledge of Nature." In Seventh International Congress of Ethnobiology, edited by John R. Stepp, Felice S. Wyndham and Rebecca K. Zarger. Athens, GA: The International Society of Ethnobiology.

UNDP. 2003. Human Development Report 2003: Millennium Development Goals: A compact among nations to end human poverty. United Nations Development Programme.

_____. 2011. The Human Development Report 2011. Sustainability and Equity: A better future for all. United Nations Development Programme.

Vandebroek, I., P. Van Damme, L. Van Puyvelde, S. Arrazola, and N. De Kimpe. 2004. "A comparison of traditional healers' medicinal plant knowledge in the Bolivian Andes and Amazon." Social Science and Medicine 59(4):837-49.

van der Ploeg, Jan D. 2008. The new peasantries: Struggles for autonomy and sustainability in an era of empire and globalization. London: Earthscan.

Vázquez G., J. Antonio , and Ramón Cuevas G. 1995. "Fotografía y Vegetación de la Sierra de Manantlán, Jalisco-Colima, México." In Flora de Manantlan: Plantas vasculares de la Reserva de la Biosfera Sierra de Manantlan, edited by J. Antonio Vazquez G., Ramon Cuevaz G., Theodore S. Cochrane, Hugh H. Iltis, Francisco Santana M. and Luis Guzman H., 50-60. Autlan, Jalisco, Mexico: SIDA, Botanical Miscellany.

Von Hagen, Victor Wolfgang , and Alberto Beltrán. 1961. The Ancient Sun Kingdoms of the Americas. Cleveland, Ohio: World Publishing.

Wallerstein, Nina. 2006. "Commentary: challenges for the field in overcoming disparities through a CBPR approach." Ethnicity and Disease 16(Supplement):S146-8.

Wallerstein, Nina, and Bonnie Duran. 2006. "Using community-based participatory research to address health disparities." Health Promotion Practice 7(3):312-23.

Wallerstein, Nina, and Edward Bernstein. 1994. "Introduction to community empowerment, participatory education, and health." Health Education Quarterly 21(2):141-8.

Walsh Sanderson, Susan R. 1984. Land Reform in Mexico: 1910-1980. New York: Academic Press, Inc.

Ward, Peter. 1986. Welfare Politics in Mexico: Papering over the cracks. London: Allen & Unwin.

Weber, Max. [1904] 1949. "Objectivity in Social Science and Social Policy." The Methodology of the Social Sciences:50–112.

Weller, Susan C., and A. Kimball Romney. 1988. Systematic data collection. Newbury Park, Calif.: Sage University Press.

West, Paige. 2006. Conservation is our government now: The politics of ecology in Papua New Guinea. Durham, N.C.: Duke University Press.

West, Paige, James Igoe, and Daniel Brockington. 2006. "Parks and Peoples: The Social Impact of Protected Areas." Annual Review of Anthropology 35:251-277.

Wheeller, Brian. 1991. "Tourism's troubled times: responsible tourism is not the answer." Tourism Management 12(2):91-96.

_____. 1995. "Egotourism, sustainable tourism and the environment - a symbiotic, symbolic or shambolic relationship?" In Tourism and the State of the Art, edited by A.V. Seaton, 647-654. Brisbane: John Wiley & Sons.

WHO. 1998. World Health Day: Safe Motherhood, 7 April 1998. Division of Reproductive Health (Technical Support): World Health Organization.

_____. 2004. Fact Sheet: Making Pregnancy Safer. http://www.who.org: World Health Organization.

Wilshusen, Peter R., Steven R. Brechin, Crystal L. Fortwangler, and Patrick C. West. 2002. "Reinventing a square wheel: Critique of a resurgent "protection paradigm" in international biodiversity conservation." Society and Natural Resources 15:17-40.

Wilson, Paul N., and Gary D. Thompson. 1993. "Common Property and Uncertainty: Compensating Coalitions by Mexico's Pastoral 'Ejidatarios'." Economic Development and Cultural Change 41(2):299-318.

Wolf, Eric R. 1958. "The Virgin of Guadalupe: A Mexican National Symbol." The Journal of American Folklore 71(279):34-39.

World Bank. 2005. WDI Database: HNPstats Data Query. Geneva: World Bank.

Zarger, Rebecca K. 2002. "Acquisition and Transmission of Subsistence Knowledge by Q'eqchi' Maya in Belize." In Seventh International Congress of Ethnobiology, edited by John R. Stepp, Felice S. Wyndham and Rebecca K. Zarger. Athens, GA: The International Society of Ethnobiology.

Zarger, Rebecca K., and John R. Stepp. 2004. "Persistence of Botanical Knowledge among Tzeltal Maya Children." Current Anthropology 45(3):413-418.

Zent, Stanford. 2001. "Acculturation and ethnobotanical knowledge loss among the Pairoa of Venezuela: demonstration of a quantitative method for the empirical study of TEK change." In On biocultural diversity: linking language, knowledge, and the environment, edited by Luisa Maffi, 190-211. Washington, D.C.: Smithsonian Institute Press.

Index

accountability, 84
acculturation, 64–65
adventure tourism, *92*
agrarian reform, 27, 31, 75
agricultural plots. *See parcelas*
agriculture, 6, 14, 68, 133; census on, 12; as economic activity, 45, 46–49, *47–48*; permanent land for, 21; productivity, 52; surface area for, 52
agroecosystems, 49–51, 68, 153
agroforestry, 68–69, 83
Ahmed, Selena, 63, 65–68
Alcorn, Janis B., 49
alloparenting, 36, 149
allopathic medicine, 125
Amazon, 62–63, 66
animal husbandry, 6, 14, 134, 149; census on, 12; as economic activity, 45, *47–48*, 49
artesanias (artisanal crafts), 99
Article 27, 31
assimilation, xvi
Ávila Camacho, Manuel, 28
Ayotitlán, 10, *81*; agriculture in, 46–48; animal husbandry in, 49; biodiversity in, 23–24; La Casa in, 121, 139–47; celebrations in, 38–39; community dynamics in, 79–85; community meetings in, 72; community structures and profiles in, 73–75; Council of Elders in, 18–19; health profile in, 112–13; imperialism in, 9; livelihood in, 133; local perceptions in, 54; material culture in, 37–38; MPK in, 130, 132–33; natural resource usage in, 45; sampling strategy in, 11–12; study sample in, 32–33, 35–46; traditional healers in, 120–24
Ayuguila River, 53
Aztecs (indigenous group), 41n1

banda (regional music genre), 40
Benz, Bruce, 9, 26, 62, 65, 127, 132
biocultural theory of human needs and values, 93
biodiversity, xvii, 48, 52, 58, 99, 137; in SMBR, 22–24; TEK and, 67–69. *See also* Manantlán Institute of Ecology and Conservation of Biodiversity
biomedical practitioners, 111–12
biomedicine, 120, 125, 140
Björk, Peter, 96, 107
black markets, 43
blending of races. *See Mestizaje*
Boas, Franz, 70n2
Bolivia, 61, 62–63

Brechin, Steven R., 83, 85, 88

Cabañas, 101–2, *106*
Café de Cuzalapa, 19
Calderón, Felipe, 28
calpulli, 25–26, 149
Canclini, Nestor, 109
Cárdenas, Lázaro, 28
La Casa de Medicina Tradicional (Traditional Medicine House), 121, 139–47
Case Western Reserve, 10
cash payments, 68
Catholicism, 7, 28, 39–41, 115
cattle ranching, 14, 49, 129, 134, 147
CDI. *See* National Commission of the Development of Indigenous Peoples
Ceballos-Lascuráin, Héctor, 96–97
celebrations, 26, 38–41
census, 12; data, 32, *34*, 74; inventories, 15
childbirths, 112, *113*
child health, 112–13
chronic diseases, 114
coding errors, 18
coffee, 99–101
cohort effects, 62
Colima, 21–22, 75, 78, 110
Colima Railway, 53
colonial mentality, 105
colonization period, 26–27
Color de la Tierra (Color of the Earth), 99–101, 103–4, *106*
co-management, 91
Commission of Communal Goods, 73
commoditization, of natural resource knowledge, 69
communal grazing areas, 52–53
community, 41n4, *45*; benefits of, 19–20; capacity building, 52, 77, 149; change in, 62; development projects of, 19; dynamics, 78–85; land ownership, 25; leadership, 76; meetings, 72–73; profiles of, 73–75; reciprocating with, 18–20; structures of, 73–75
community-based sustainable development, 72, 83, 87
community-based tourism, 91, 96
competition, 95
complexity, 145–46
comunidades indígenas (indigenous communities), 27, 30–32, 73, 147
CONANP. *See* National Commission of Protected Areas
Conference of Indigenous Communities (*Convocatoria de Comunidades Indigenas*), 37
conquest period, 26–27
Consejo de Mayores. See Council of Elders
conservation, 1, 7, 31; of biodiversity, 67–69; development programs and, 87–89; economic activities and, 45–49; governance and, 53–55, 71–72; integrated, 89–102; local perceptions and, 53–55; natural resource usage and, 43–45; programming, 49–54. *See also* Integrated Conservation and Development Projects; Manantlán Institute of Ecology and Conservation of Biodiversity

Conservation Program for Sustainable Development (PROCODES), 29, 31, 101
Constitution of Mexico (1917), 27
consumption patterns, 74
contamination, of rivers and streams, 43
contemporary period, 27–30
Convocatoria de Comunidades Indigenas. *See* Conference of Indigenous Communities
cooperatives, 100, 123
corpus, 59
Council of Elders, 18–19, 79, 82, 84–85; La Casa and, 143, 147; celebrations and, 38–39; indigenous revitalization movement and, 73
Council of Members, 27
crop rotating system, 83
CUCSUR. *See* South Coast University Campus of the University of Guadalajara
culture: acculturation, 64–65; brokers, 89; consensus model, 60, 62–63; domains of, 128–30; heritage of, 2, 98, 104, 124, 137, 140; imperialism, 9; interconnectedness of, 136; intracultural heterogeneity, 50; knowledge of, xvii, 2–3, 137; local identity of, 37–41; material, 37–38, 121, 123; polyculture, 83; traditional, 99; traditions of, 7
curanderos, 112, 114, 121, 140, 149
Cushing, Frank Hamilton, 70n2
Cuzalapa, 26, 98–101, 103–5

Davis, Benjamin, 35
deciduous evergreen forests, 22
decision-making, 79–80, 82, 84; agrarian reform and, 30; governance and, 72; health decisions, 144, 146; identity and, 109; land tenure systems and, 27; local-level processes of, 89; outside intervention and, 76
demographics, 13, *33*
descriptive data, *34*
development, 1, 9, 63, 128; assessments of, 142; critiques of, 72; economic, 2–3, 67; social, 2. *See also* sustainable development
development programs: integrated conservation and, 89–102; sustainable development and conservation, 87–89
Dewey, John, 93–94
dichotomous variables, 137n2
differentiation, xvi
distribution: of cultural knowledge, 137; of study sample data, *11*
domestic violence, 115
drugs, illegal, 43
Dry Laws, 40

Ecological Knowledge Score, 63
ecologically noble savage, 67
ecological studies, 77
economic activities, 14, 35, 45–49, 134; agriculture, 45, 46–49, *47–48*; animal husbandry, 45, *47–48*, 49; profile, 5
economic changes, 2, 63
economic development, 2–3, 67
economic information, methods for, 17–18

economic models, 93–96
ecosystems, 49
ecotourism, 19, 39, 53, 87, 91–93, *92*; analytical framework for, 102–7; discourse of, 93–96; ethnography of, 98–102; in Latin America, 96–98
Ecuador, 66
education, 33; environmental, 96; PROGRESA, 75; UNESCO, 50
Ejidal Assembly, 84
ejidatarios, 27, 150
ejidos, 10, 27, 30–32, 73, 147, 150
emic, 94–95, 107, 150
encomienda, 26, 27, 150
environmental education, 96
environmental policies, 94
ethnobiology, 58
ethnobotanical knowledge, 127
ethnoecology, 58; quantitative approaches in, 59–61
ethnomedical knowledge, 3
ethnomedical systems, 7, 150; biomedical practitioners, 111–12; healthcare system, 109–11; medical pluralism, 114–16; traditional healers, 116–24
etic, 94–95
evergreen-mesophytic mountainous forests, 22
ex situ, 51, 145, 151
EZLN. *See* Zapatista National Liberation Army

federal clinics, 110
federal forest rangers, 20n1
field notes, 16
fieldwork, 8–20; data quality, 17–18; orientations to Reserve, 9–11; preliminary preparation, 9–11; reciprocating with study communities, 18–20; sampling strategy, 11–17
field workers, 54
Finkler, Kaya, 114–15
Fletcher, Robert, 95–96, 104
folk medicine, 114, 151
folk taxonomy, 63, 67
food plants, 123
forests, 22–23, *23*; agroforestry, 68–69, 83; federal rangers, 20n1; management of, 44–45, 75, 87
formal rationality, 93–94
Fox, Vicente, 28
freelists, 15, 18, 127, 151; cultural domains and, 129–30; folk taxonomy and, 63

Garro, Linda C., 61–62
gender, 61–62, 65–66, 115, 132
geographic area, 121–22
geotourism, 91, *92*, 96
Gerritsen, Peter R. W., 86n2, 137
globalization, xvii, 1–2, 69, 105, 147, 151; cultural interconnectedness and, 136; health behaviors and, 115; TEK and, 59
global worldviews, 2
Godoy, Ricardo, 61, 63, 64, 66
governance, 53–55, 71–72, 144; community dynamics and, 78–85; community meetings and, 72–73; community structures and profiles, 73–75; outside intervention and, 76–78
government: agencies, 99, 141; welfare programs, 53
grazing areas, communal, 52–53
Green Party, 28
green tourism, 91, 96

El Grullo, 23
Guest, Greg, 66
guided hikes, 101

haciendas, 26, 151
Haenn, Nora, 104
healers, traditional, 112, 116–24
healthcare: childbirth services, *113*; privatization of, 28; providers, 36; system, 109–11
health decisions, 144, 146
Health for All. *See Salud Para Todos*
health information, methods for, 17–18
health profile, 7, 112–13
heritage tourism, 91, 96
Highman, James E. S., 96
hikes, 101
Holy Week. *See Semana Santa*
Honduras, 61, 66
household characteristics, 14
human dignity, 83, 88
hybridization, 91

ICDPs. *See* Integrated Conservation and Development Projects
ideal rationality, 102
identity: cultural, 37–41; formation of, 109
illegal drugs, 43
illiteracy, 33, 74
IMECBIO. *See* Manantlán Institute of Ecology and Conservation of Biodiversity
IMSS. *See* Mexican Institute of Social Security
Independence Day, 38
indices, 130, *131*; relationships between, 133–35
indigenous communities. *See comunidades indígenas*
indigenous knowledge, 15, 70n2, 91, 135, 139, 145
indigenous-revitalization movements, 65, 73
indigenous technology, 58, 135
industry, 99, 141
INEGI. *See* National Institute of Statistics and Geography
infant mortality rates, 113
injuries, 114
in situ, 51, 151
Institutional Revolutionary Party (PRI), 28
institutions, 76
integrated conservation, 89–102
Integrated Conservation and Development Projects (ICDPs), 75, 85, 90, 92–93; conservation programming and, 50–51; decision-making and, 80; natural resource management and, 72; stakeholder institutions and, 77
intensive agriculture techniques, 52
intracultural heterogeneity, 50

Jalisco, 21
Jalisco-Colima volcanoes, 22
jaripeo (rodeos), 41
Las Joyas (research station), 54

Kapp, Karl William, 93–94
Kleinman, Arthur, 116
knowledge: cultural, xvii, 2–3, 137; ethnobotanical, 127; ethnomedical, 3; generational loss of, 65; indigenous, 15, 70n2, 91, 135, 139, 145; practical, 59; scientific, 51, 91; theoretical, 59–60; transmission of, 58. *See also* medicinal plant

knowledge; traditional ecological knowledge
kosmos, 59

labor, 46, 75; divisions of, 61
Laitner, Karen, 26
land ownership, community, 25
land tenure systems, 26–27, 46, 73, 75, 147
Land Use Importance Value, 63
language, 26, 41n1, 65
large-scale industry, 88
legitimacy, 84, 88
life history interviews, 12
literacy, 29, 127
livelihood, 1, 4, 14, 133
livestock, 50, 75
local cultural identity: material culture, 37–38; tradition and celebration, 38–41
local markets, 136
local political structures, 30
lumber companies, 53

MAB. *See* Man and the Biosphere
maize, 25, 44, 83
Management of the Biosphere Reserve, 9, 20n1, 44–46, 52–54, 73, 104–6, 142; *Color de la Tierra* and, 99–100; community dynamics and, 80; political action and, 85; sustainable development and, 77
Man and the Biosphere (MAB), 21, 50
Manantlán Institute of Ecology and Conservation of Biodiversity (IMECBIO), 9, 11, 18, 85; conservation programming and, 51–52; forests and, 23; governance and, 54; sustainable development and, 37
Mann-Whitney test, 132
Manzanillo, 22
maps: of main study communities, *45*; national, *22*; regional, *24*
marijuana, 43
market integration, 58, 63–65, 128; biodiversity conservation and, 67–69; TEK and, 65–69
market participation, 63–65, 152
mass tourism, *92*
material culture, 37–38, 121, 123
maternal health, 112–13
McGehee, Nancy Gard, 103
Meares, Alison C., 103
medical pluralism, 7, 109, 114–16
medicinal plant knowledge (MPK), 1–5, 116, 121, 125, 127–30, 135–36, 152; indices and, 130, *131*, 133–35; livelihoods and, 13; socioeconomic status and, 133; variation in, 132–33
medicinal plants, 38, 109, 114, 116–22
Mestizaje (blending of races), 144
Mexican Institute of Social Security (IMSS), 28–29, 109–10, 112
Mexican model, 44, 141
Mexican Revolution (1910-1915), 27
midwives, 111–12
migration, 35, 115
migratory waged labor, 46
Miguel de la Madrid, 44
milpa (shifting maize cultivation), 25, 49, 152
Minatitlan, 22

Ministry of Health, 29, 109–11
modernization, 2, 9, 74, 82, 128, 134, 142; acculturation and, 65; community change and, 62; market integration and, 63
modern technology, xvi
monocropping, 50, 52
MPK. *See* medicinal plant knowledge
music, 40

NAFTA. *See* North American Free Trade Agreement
Nahua (indigenous group), 6, 21, 25–26, 41n1, 80, 140
Nahuatl (language), 26, 41n1
Napolitano, Valentina, 115–16
National Action Party (PAN), 28
National Commission of Protected Areas (CONANP), 101
National Commission of the Development of Indigenous Peoples (CDI), 37, 78
National Development Plan (1995-2000), 29
National Institute of Statistics and Geography (INEGI), 12, 32
national map, 22
national migration, 35
national social movements, 75
natural resource management, 7, 31, 68, 83
natural resource usage, 43–45
nature-based tourism, 91, *92*, 96
naturistas, 112, 140, 152
neoliberalism, 103, 105; markets, 87; strategies, 95
Nichter, Mark, 89
non-farming activities, 129
North American Free Trade Agreement (NAFTA), 28, 152
nurses, 111

objectives, of study, *5*
Oldfield, Margery, 49
Oportunidades, 75, 122, 143
outside intervention, 76–78
outside markets, 35

paid labor, 46, 75
PAN. *See* National Action Party
parcelas (agricultural plots), 27, 153
parteras (midwives), 111–12
participant observation, 9, 61
participatory approaches, 91
participatory tourism, 91
paternalistic models, of resource management, 136
Peña Nieto, Enrique, 28
personal health histories, 14–15
Peru, 95
Pfeiffer, James, 89
pharmaceuticals, 119
physicians, 111
pile sorts, 63
pine forests, 22
pluralism, 7, 109, 114–16, 144–45
political-economic systems, 57
political processes, 82
political structures, local, 30
pollution, 53
polyculture, 83
popular medicine, 114, 153
population growth, 29–30
Porfirio Diaz, José de la Cruz, 26–27
poverty alleviation, 95
practical knowledge, 59
Pragmatic Instrumentalism, 94
praxis, 59

preliminary preparation, 9–11
PRI. *See* Institutional Revolutionary Party
primary surveys, 13
private industry, 141
private practitioners, 110
privatization, 142; of healthcare, 28; of land, 31; of resource ownership, 95
PROCODES. *See* Conservation Program for Sustainable Development
PRODERS. *See Programa de Desarrollo Regional Sustentable*
productivity, agricultural, 52
professional medicine, 114, 153
Programa de Desarrollo Regional Sustentable (PRODERS), 101
Program for Education, Health and Nutrition (PROGRESA), 75
PROGRESA. *See* Program for Education, Health and Nutrition
pro-poor tourism, 91, *92*, 96
protected areas, 1, 7, 68, 93, 153
public health, 30; campaigns for, 14

qualitative methods, 16–17
quantitative methods: in ethnoecology, 59–61; in sample strategy, 13–15
questionnaires, 13

rational humanism, 93
rationality, 99; formal, 93–94; ideal, 102; substantive, 93–94, 104
regional offices, 27
religious syncretism, 25
reproductive health, 112–13
research: groups, 99, 141; institutions, 76, 141; questions, 1–3
resistance movements, 65
responsible tourism, *92*
Reyes-Garcia, Victoria, 59, 62, 66
ritual traditions, 25
rodeos, 41
rural sustainable development, 95
rural-to-international migration, 35
rural-to-rural migration, 35
rural-to-urban migration, 35

Salinas de Gortari, Carlos, 28, 31
Salud Para Todos (Health for All), 124
sampling strategy, 11–17; qualitative methods, 16–17; quantitative methods, 13–15
scientific knowledge, 51, 91
secondary activities, 10
secondary stakeholders, 71, 73
Secretary of Environment and Natural Resources (SEMARNAT), 78
seeds, 48, 50
Semana Santa (Holy Week), 38
SEMARNAT. *See* Secretary of Environment and Natural Resources
semi-structured interviewees, *117*
shifting maize cultivation. *See milpa*
Sierra of Manantlán Biosphere Reserve (SMBR), 1, 6, 9, 65, 75; biodiversity in, 22–24; colonization of, 26–27; *comunidades indígenas* in, 30–32; conquest of, 26–27;

contemporary period of, 27–30; *ejidos* in, 30–32; geographic area of, 121–22; Nahua in, 25–26; national map of, *22*; outside intervention in, 76–78
social action, 87
social costs, 94
social development, 2
social movements, 75, 80, 89
social networks, 35
social relationships, *36*
Social Security administered clinics, 110
social values, 93–96
social welfare programs, 35, 75, 82, 122, 143
socioeconomic indices, 130, *131*; relationships between, 133–35
socioeconomic profile, 128
socioeconomic status, 32–33, 89, 115, 133
sociopolitical factors, 112
South Coast University Campus of the University of Guadalajara (CUCSUR), 9–10, 17, 38–39, 51, 77, 122–23
specialized institutes, 110
SPSS. *See* Statistical Package for the Social Sciences
stakeholders, 71, 75–76, *76*, 79, 99, 141
Stakeholder Theory, 95
state-based regional offices, 27
Statistical Package for the Social Sciences (SPSS), 18
Steward, Julian, 57
Stronza, Amanda, 95
structural adjustment policies, 28
study: ecological, 77; main communities of, *45*; main hypothesis of, *5*; model for, *4*, 4–6; objectives of, *5*; of TEK, 59–63; variables of, 3–4. *See also* fieldwork
study sample: census data of, *34*; demographics of, *33*; description of, 32–37; descriptive data of, *34*
subsistence strategies, 24; conservation programming and, 49–54; economic activities and, 45–49; governance and, 53–55; local perceptions and, 53–55; natural resource usage and, 43–45
substantive rationality, 93–94, 104
sustainable development, xvii, 6, 44, 52, 77, 83–84; community-based, 72, 83, 87; development programs and, 87–89; essential goal of, 90; governance and, 71–72; IMECBIO and, 37; integrated conservation and, 89–92; PROCODES, 29, 31, 101; rural, 95
sustainable tourism, 91, *92*, 96

tea, 66, 68
technology: indigenous, 58, 135; modern, xvi
TEK. *See* traditional ecological knowledge
Telcruz, 10; agriculture in, 46–48; animal husbandry in, 49; biodiversity in, 23–24; health profile in, 112; livelihood in, 133; material culture in, 37; MPK in, 128, 130, 132–33; sampling strategy in, 11–12; study sample in, 32–34; TEK in, 62
tenure systems, 26–27, 46, 73, 75, 147
teocintle, 21, 44

El Terrero, 98–99, 101–2, 103–5
theoretical knowledge, 59–60
Theory of Social Costs, 94
Tlaloc (god of water), 25
Toledo, Victor M., 59–60
Tonantzin (goddess of earth), 40
tourism, 91, *92*. *See also* ecotourism
trade-offs, 142, 143, *145*
trade systems, 50
tradition, 38–41, 74
traditional agroecosystem, 49–51, 129, 153
traditional culture, 99
traditional ecological knowledge (TEK), 3, 57–59, 134, 136; biodiversity conservation and, 67–69; future directions for, 69–70; market integration and, 65–69; studying, 59–63
traditional healers, 112, 116; in Ayotitlán, 120–24; in Zacualpan, 117–20
Traditional Medicine House. *See La Casa de Medicina Tradicional*
traditional plant knowledge, 2
training, medical, 111
transgenic seeds, 48–49
transnational migration, 35
tropical deciduous forests, 22–23
Tsimane' Amazonian Panel Study, 63

underground black markets, 43
United Nations Educational, Scientific and Cultural Organization (UNESCO), 50
University of Guadalajara, South Coast (CUCSUR), 9–10, 17, 38–39, 51, 77, 122–23
urban centers, 23–24, *24*
urbanization, 30

values, 65; social, 93–96
Vandebroek, I. P., 62
variables, of study, 3–4
vegetation, 22–23
Virgin of Guadalupe, 25, 39–41
volcanoes, 22

waged labor, 75; migratory, 46
Wallerstein, Nina, 72
Weber, Max, 93–95, 98–99, 102–3
welfare programs, 35, 53, 75, 82, 122, 143
Wilshusen, Peter R., 55n2
win-win approaches, 8, 140–41, 142
World Health Organization, 111
worldviews, global, 2

Zacualpan, 10, 18, *81*, 85; agriculture in, 46–48; animal husbandry in, 49; biodiversity in, 23–24; celebrations in, 41; community dynamics in, 78–79, 80–83; community meetings in, 72; community structures and profiles in, 73–75; health profile in, 112–13; livelihood in, 133; local perceptions in, 54; material culture in, 37; MPK in, 130, 132–33; natural resource usage in, 45; sampling strategy in, 11–12; study sample in, 32–35; traditional healers in, 117–20

Zapatista National Liberation Army (EZLN), 31, 55, 79–80, 89
Zea perennis, 21
Zedillo, Ernesto, 28–29

About the Author

Elizabeth Anne Olson is Assistant Professor of Global Health Studies and Environmental Science at Allegheny College in Meadville, Pennsylvania. She received her doctorate of philosophy in Anthropology from Case Western Reserve University in Cleveland, Ohio. She is both a medical and an environmental anthropologist whose research has looked at traditional healing systems and the relationships between humans and our environments in the United States, the Bolivian Amazon, Mexico, and Western Europe. Professor Olson's ongoing research in Mexico encompasses sustainable development, health, and ethnobiology.

By virtue of her work and research, she has found many opportunities to be involved with various social justice and community-based conservation initiatives. In her spare time, she enjoys competing in marathons and birding—both in Mexico and across the United States—with her son Gabriel.

Lightning Source UK Ltd.
Milton Keynes UK
UKOW05n0106110717
305090UK00006B/43/P

9 780739 176634